T0211338

Determinism and Free Will

Fabio Scardigli · Gerard 't Hooft
Emanuele Severino · Piero Coda

Determinism and Free Will

New Insights from Physics,
Philosophy, and Theology

Springer

Fabio Scardigli
Department of Mathematics
Politecnico of Milano
Milano, Italy

Gerard 't Hooft
Institute for Theoretical Physics
Utrecht University
Utrecht, The Netherlands

Emanuele Severino
Brescia, Italy

Piero Coda
Istituto Universitario Sophia
Firenze, Italy

ISBN 978-3-030-05504-2   ISBN 978-3-030-05505-9   (eBook)
https://doi.org/10.1007/978-3-030-05505-9

Library of Congress Control Number: 2018965905

This Springer imprint is published by the registered company Springer Nature Switzerland AG
The registered company address is: Gewerbestrasse 11, 6330 Cham, Switzerland

Contents

Introduction

This book is a collection of three essays, written by theoretical physicist Gerard 't Hooft, philosopher Emanuele
Severino, and theologian Piero Coda, and inspired by
the talks the three authors made as keynote speakers at
the conference "Determinism and Free Will", held at the
Cariplo Foundation Congress Center in Milan on May
13, 2017.

The conference was conceived and organized by a group
of friends and colleagues consisting of Fabio Scardigli,
Marcello Esposito, and Marco Dotti. We are grateful to
our colleague Massimo Caccia, and especially to colleagues
Gabriele Gionti and Massimo Blasone, for their help
before and during the workshop.

The idea of organizing a meeting between Severino and
't Hooft had already been conceived several years ago. In
fact, there had been a couple of unsuccessful attempts in

This Introduction has been completed at the University of Leiden, in October
2017.

© Springer Nature Switzerland AG 2019
F. Scardigli et al., *Determinism and Free Will*,
https://doi.org/10.1007/978-3-030-05505-9_1

2010 and 2012 at the "DICE" Theoretical Physics conferences organized by Thomas Elze in Castiglioncello on a bi-annual basis. The opportunity arose this year in May, when 't Hooft was in Castelgandolfo (Rome) for a conference organized at the Vatican Observatory and, on the way back to Holland, kindly agreed to stop for a couple of days in Milan, a city easily reachable by both Severino and Coda.

The conceptual reasons that led to this encounter lie first of all in the line of research pursued by 't Hooft for several years now, in which he aims to provide quantum mechanics with a deterministic foundation. His program seeks to bring this theory back under the umbrella of the most stringent determinism, a goal pursued by Einstein during the last decades of his life. On the other hand, Severino has built up an ontological vision that radically negates any reality in the becoming, a point of view often associated with the strict deterministic conception of reality promoted by Einstein and Spinoza. He thus seemed to be the natural philosophical interlocutor for the physicist from Utrecht. Considering then the endless interweaving of the theme of free will with so many aspects of human experience, and also the happy accident of the 500th anniversary of the thesis presented by Luther (1517–2017), it seemed appropriate to complete the trio of speakers with the theologian Coda, who has always devoted a lot of attention to these issues.

The following enumerated sections address different aspects of the debated topics.

1. In Severino's vision, "becoming" (understood as the coming out of and the return to nothing of things) does not exist, i.e., it is not an element of reality. Becoming, far from being the most obvious, trivial, and undeniable evidence of the world, is indeed

a theory, that is, just one 'interpretation' of events among the many possible. Indeed, Severino thinks, and thinks he has shown, that the interpretation of becoming, manifested since the Greek origins of Western thought as the oscillation of things between being and nothing, is just a "very stubborn illusion", a misinterpretation of events (words very similar to those with which Einstein described time in a letter to the sister of his beloved friend Michele Besso, who had just passed away). With his philosophical research, Severino thinks he has provided a foundation for the eternity of beings, the eternity of each single entity, of each single event. This vision is undeniably similar to the vision proposed in general relativity, in which all events, past, present, and future, have always coexisted and will do so forever more, remaining eternally as points on the space-time manifold. The problem for this vision comes from the very heart of the other great theoretical construction of 20th century physics, quantum physics (at least perhaps until the recent studies by 't Hooft). Here in fact this vision clashes against Heisenberg's uncertainty principle, according to which the future is not strictly determined by the present, and the present is not strictly determined by the past, because there is a non-eliminable role played by chance in generating even elementary events. Physics, at least from the days of Maxwell and Boltzmann, has long been accustomed to using probabilistic laws to describe complex events where it is reasonable to expect chance to play an important role. The novelty in the standard formulation of quantum mechanics was that even the elementary event, the absolutely simple event (think for example of a photon emitted by an electron in an atom) happens by pure chance. On the contrary, in the deterministic interpretation that 't Hooft proposes, quantum mechanics is instead brought back to the most complete, strict Einsteinian determinism. 't Hooft's vision is thus somehow close to Severino's idea of the eternity of every single event, of the non-existence of becoming (which has always been thought of by Western philosophy as the random emergence of things from nothing).

2. It should be noted that one of the main motivations of the 't Hooft program, viz., to render quantum mechanics (QM) strictly deterministic and therefore conceptually closer to general relativity (GR), is precisely the fact that, once a greater conceptual homogeneity has been obtained between QM and GR (particularly as regards the idea of time advocated by the two theories), the much sought after goal of a unified theory of all physical phenomena would certainly be brought closer.

Indeed, if such a formulation exists, then QM will have a structure somehow similar to that of a classical theory, so it could be more easily re-formulated within the framework of general relativity. The unification of QM and GR would then in principle be far simpler and more natural.

The possibility, as shown by 't Hooft, of describing a system as simple as a cellular automaton, a perfectly classical and deterministic system, within the language of quantum mechanics, inevitably suggests that even the much more complicated system we observe, the physical world, so well described by that sophisticated quantum field theory called the Standard Model, may in fact be nothing but a very complicated but deterministic cellular automaton.

3. It has been said by several scholars that Severino's ontological vision appears to be an "influential

metaphysics" of general relativity (to use Popper's locution), a sort of "general relativity" pushed to the extreme, with consistency and rigor. Severino seems in some respects stricter than Einstein when he establishes the eternity of every being. This vision naturally fills the "spaces" (social, psychological, economic, etc.) left necessarily empty by physical theory, and the scenario is undoubtedly suggestive. As will be seen from his essay, Severino does not like to push the analogy too far between his position and the vision proposed by general relativity. In particular, he strongly emphasizes the different conceptual origins of the two logical structures. However, it must be said that the common features and the intrinsic coherence make it tempting to overlook the different origins of the two pictures. On the other hand, the scope and the terms used differ so much between them that the existence of a channel of communication between the two structures appears to be almost miraculous.

4. A possible critical point in the Severinian construction is his concept of "mathematical model" of the world. Severino says that, from its birth with Galileo and Newton, and until the end of the eighteenth century, modern science had an absolute, epistemic conception of the truth. Then, with the invention of ideas like non-Euclidean geometry and abstract algebra in mathematics, and with the quantum and relativistic revolutions in physics, the epistemic character disappeared, leaving room for the idea of science as hypothetical knowledge, designed to produce effective, working, and replaceable mathematical models of the world. This may suggest that the idea of "falsifiable" (mathematical) models appeared in physics and the other sciences only in the last two centuries. However, this view is not really corroborated by the most recent

historical reconstructions. Lucio Russo and other scholars have shown with an abundance of detail that the concept of a mathematical model of a physical phenomenon was already developed with great clarity and effectiveness by the Alexandrians, and in particular Archimedes, Eratosthenes, Heron, and others.

So the idea that scientific theories are not absolute truths, i.e., not epistemic, but instead are (more or less efficient) mathematical models of the world has been in circulation for a much longer period than the last two centuries. Following the Alexandrians, the scientists of 16th and 17th centuries (starting with Galileo, but not forgetting Leonardo da Vinci and others) subjected different physical theories (in the sense of models) to experimental tests in order to find out which one was the best, i.e., the most effective for representing the physical world. They didn't consider themselves to be hunting for absolute truth, but merely for a physical theory that was more effective than the old Aristotelian models. Hence, Newton set up his bucket experiment to test his idea of absolute space. From the mid-17th century and during the 18th century, two different mathematical models of light, a wave theory and a corpuscular theory, were in fierce competition, with neither gaining the upper hand, until finally an experiment appeared to decide the issue (Young's double slit experiment).

It is therefore legitimate to state that falsifiable conceptions of scientific truth have always been present since the Alexandrian origins of modern science, or at least that they have long coexisted with the epistemic conceptions of scientific truth.

5. The task of understanding the word "being" has long been believed to be the prerogative of pure theoretical philosophers. But today, in a world increasingly

filled with the technological products of "quantum physics", it is almost unavoidable to ask "what lies behind this", perhaps with some kind of return to the origins, to the philosophy of nature put forward by Thales, Anaximander, Democritus, or Parmenides. The old quarrels between the founding fathers of quantum physics, which many physicists long considered a pastime for the "elderly" and irrelevant to the state of the art in physics, are once again being vigorously discussed. Physicists return to what they always should have been: philosophers of nature. Fundamental questions are being raised again, as when Einstein asked an astonished Abrahm Pais at the Princeton campus: "But are you really convinced that the Moon only exists if you look at it?" Of course, the underlying motivations are very often also prosaic and concrete. For instance, one conception rather than another of the quantum state, or of the problem of measurement, can completely change a whole line of research on quantum computers, and with it, the use of millions of dollars of funding.

6. With regard to QM, 't Hooft joins a long line of outstanding physicists who have shown discomfort with the standard interpretation of QM, or even criticized its foundations. It is well known that at least two of the founding fathers of QM, Einstein and Schroedinger, put forward critical insights into various aspects of the quantum point of view. And although they have generated research for about 80 years, many aspects of those problems remain without a shared consensus among the scientific community. Let us recall here just a few of these points:

   (a)   For Einstein, QM is not a theory about single events. By definition, the fact that the theory has such a radically statistical structure prevents predictions about individual events (except for certain special cases): "The wave function psi does not describe, in any way, the condition of 'a' single system" (A. Einstein, Physics and Reality, 1936).

(b) In the famous EPR article (1935), Einstein claims to have demonstrated QM's "incompleteness": there have to be elements of physical reality that are not described, or captured, by the QM wave function.

(c) Along the same lines, in the same year, Schroedinger launches another important idea in the form of his famous "cat paradox". If we follow the standard interpretation of QM, in fact, before a direct observation ('measure') has been done, the cat should be considered both alive and dead at the same time! Just as the radioactive atom (which controls the life of the feline through a clever mechanism) would result in a linear superposition of the decayed and non-decayed states.

(d) For both Einstein and Schroedinger, the statistical character of QM, although it captures a description of the reality with which each future model must be compared, is not a good foundation upon which to build a theory able to describe single events, rather than just statistical descriptions of sets of events. Just as, according to Einstein, "the Newtonian laws of point particle mechanics could not be deduced from thermodynamics" (Physics and Reality, 1936).

Einstein and Schroedinger's attitude towards QM is what the young Einstein, influenced by Mach, expressed with regard to the fundamental concepts of absolute space and time elaborated by Newton: "The prodigious success of his doctrine [Newtonian mechanics] obscured the critical investigation of its foundations [for two centuries]"(Herbert Spencer Lecture, Oxford 1933).

7. An important topic of research in the foundations of quantum mechanics directly involves the concept of free will, a concept which might seem, at first sight, to be linked to very concrete legal or social problems rather than to the foundations of an abstract physical theory.

In fact, one of the most debated (and paradoxical) results of quantum research in recent years is the so-called Free Will Theorem. This proceeds roughly as follows. First, the authors, Conway and Kochen, give a formal definition of free will which makes it possible to "quantify" the degree of "free will" possessed by a particular entity. Then, they analyze a Bell-type experiment (involving electron spin or photon spin/helicity), and demonstrate that, on the basis of commonly accepted QM principles, the observed electron (photon) must have the same degree of "free will" as the observer who performs the experiment.

The paradoxical and astonishing aspect of this conclusion is evident. How could an elementary particle (elementary, therefore without structure) have the same degree of free will as the human being who observes it? The real purpose of the theorem thus appears to be to highlight the paradoxical aspects of QM, rather like the Schroedinger cat experiment, but in another context.

For some, the content of the Free Will Theorem is even tautological. Indeed, if the world is completely deterministic, then neither the electron nor the observer have any free will because everything is completely predetermined. If, on the other hand, we admit that the observer has free will, then the world is not completely deterministic, and we pay the price of seeing the electron exhibiting an indeterminacy, a "freedom" of choice, almost "its own free will".

8. Bell's inequality is still the most frequently invoked argument against the possibility of building deterministic and local models of quantum phenomena. The vast majority of physicists believe that the lengthy debate triggered by Einstein's criticism in the 1930s has been definitively closed in favor of a non-deterministic interpretation of QM since the appearance of Bell's theorem in 1964. Those who propagate a return to determinism are often viewed as people (by now) far from the mainstream of scientific research. Nevertheless, some of the most original thinkers of today, including 't Hooft, Penrose, Ghirardi, and others, have questioned various aspects of the standard Copenhagen interpretation of QM. Bell's inequality plays a key role in favor of the standard interpretation. However, the importance of the hypothesis of "measurement independence" in demonstrating the theorem was already clear to John Bell, and subsequently to other scientists like Shimony, Clauser, Horn, and others. This is an hypothesis that can be tied (and often is) to the "free will" of the observer who performs or oversees the measurement; that is to say, tied to the freedom of the observer to arbitrarily choose the orientation of the polarizing filters used in the measurement. On the decisive role played by this apparently innocent (and obvious) hypothesis, it is interesting to recall a recollection by 't Hooft himself, according to which, during a meeting some thirty years ago, John

Bell said: "If free will does not exist, then the deduction of the Bell inequalities is not valid."

In other words, the hypothesis of free will, or the observer's freedom of choice, is essential to the proof of Bell's inequalities. The latter are obeyed by any theory (with hidden variables) that is deterministic and local, and are violated by quantum mechanics. This is the standard argument that excludes a priori all local deterministic models of quantum phenomena involving hidden variables, since they do not violate Bell's inequalities, while QM does. Most people renounce deterministic local models in favor of quantum indeterminacy. However, Bell's inequality is clearly a consequence of the measurement independence hypothesis, which can in turn be naturally connected to the more than "obvious" assumption of freedom of choice for the observers themselves.

The use of the free will postulate (or equivalent assertions) to prove Bell's inequalities is confirmed also by the most recent formulations of such inequalities (see, for example, Brukner, Costa, Pikovski, Zych, "Bell Theorem for Temporal Order", arXiv:1708.00248). So, Bell's theorem and its (indirect) support for QM may appear as a kind of projection of the "obvious" hypothesis of attributing "free will" to human beings. Although it is not the only working model, QM appears instead under the strange light of being the model that fulfills our (natural) desire to attribute free will to us humans! One could almost say, in this subtle and specific sense, that QM is a "projection" of the human mind, owing to the dogma, which sounds typically Ptolemaic, of maintaining that humans possess the property of free will. These ideas fit well with those of the Free Will Theorem, whose authors claim, after giving a mathematical definition of the concept of free will, that if QM is true then the electron and its (human) observer have exactly the same degree of free will, a clearly absurd situation.

9. The "hidden" but obvious hypothesis behind Bell's inequality is that of "measurement independence", closely related to the possibility of attributing freedom of choice (or free will) to the observer who performs or oversees the measurement. Somehow, since we humans *want* to have free will, we must therefore also attribute it to elementary particles. We cannot admit a deterministic description of the micro world, otherwise we too would be deterministic and we would not have free will. From this prospective QM looks almost like a "choice". Humans want to have free will, so they naturally have to choose QM (which somehow guarantees it) over and above other models, which are discarded even though they could work (such as Bohmian mechanics, for example, at least in the non-relativistic regime), essentially *because* they are deterministic (and non-local).

Of course, the Severinian aspect of this situation will not have escaped the reader: we want, we believe, we choose to have free will. In a sense, we "choose" the world to be indeterminate to preserve our supposed free will; we somehow "choose" a world that is "becoming" (indeterminism) in order to better manipulate it. In this above-mentioned sense, the usual non-deterministic interpretation of QM looks rather like a "projection" of our mind.

The prevalence of a non-deterministic vision in the standard interpretation of quantum mechanics has been described by Severino in his book "Law and Chance" as a result of the more general course of Western philosophical thinking over the last two centuries. In Severino's words, "willpower 'wants' 'becoming' to exist, wants things to come out of nothing without a cause (randomly), to maximize the possibility of manipulating them". In some way, it wants standard QM to be the only proper representation of the physical world.

10. The centrality of the hypothesis of freedom of choice for the observer is also emphasized by other authors. For example, Hossenfelder, in her blog, points out that if we deny the "free will" hypothesis, we lose Bell's theorem:

"The free will of the observer is a relevant ingredient in the interpretation of quantum mechanics. Without free will, Bell's theorem does not hold, and all we have learned from it goes out the window."

"The option of giving up free will in quantum mechanics goes under the name of 'superdeterminism' and is extremely unpopular. Unfortunately, it is highly probable that by scorning 'superdeterminism' we will miss something really important, something that could very well be a basis for future technologies."

"This kind of theories are often called 'conspiracy' theories, as it seems that the universe must be deliberately meant to prevent experimentalists from doing what they want. Therefore, this option is often not taken seriously."

"However, this could be a misleading interpretation of 'superdeterminism'. All that 'superdeterminism' means is that a state cannot be prepared independently of the detector settings. That doesn't necessarily imply a 'spooky' action at a distance, because the backwards light cones of the detector and state (in any reasonable universe) intersect anyway."

Of course, experiments have been planned (and some have already been done) that use very distant objects to close the loophole of "free will" or "measurement independence": stellar light experiments (Zeilinger, January 2017), and even with light emitted from quasars billions of light-years away (Kaiser, Gallicchio, planned for 2017–18). However it is clear that it is not possible to completely exclude an intersection between the particle light cone and that of the measuring instrument in the remote past. After all, it is assumed they both originate in the big bang.

11. If we adopt the superdeterminist perspective, there is a problem in justifying the apparent, actual presence of "free will", of the ability to make "free" choices, that each of us experiences in everyday life. On this point it is interesting to report the ideas of Seth Lloyd, who shows that the illusion of "having" free will comes from the computational complexity of the processes (decision-making) that take place in our minds. A system, fundamentally deterministic but complex, is not able to predict its own decisions before taking them, because anticipating them has a degree of complexity similar to, or greater than, what is necessary to actually take them, and put them into practice. And this total blindness with regard to its own future choices, translates for the given system into the illusion of having freedom of choice, of exercising free will.

In fact, predicting ("calculating") what decision you will take in 10 minutes from now is such a complex and lengthy process, at least as complex and lengthy as actually taking it. From here arises the illusion of freely deciding.

Free will reflects our ignorance of ourselves. Lloyd argued that free will is meaningful because our own decisions are unknown to us until we make them. Like all great theorems in computer science, his argument appeals to the paradoxes of recursion. When you think about yourself, you think about thinking about yourself, and you enter an endless loop: "Any system that can ask what it will be doing in five minutes' time, cannot always answer it. [...] It is less effective to simulate yourself, than it is simply to be yourself."

Lloyd summarizes this point of view on free will in a paper published in 2012, entitled "A Turing test for free will": "This article investigates the roles of quantum mechanics and computation in free will. Although quantum mechanics implies that events are intrinsically unpredictable, the pure 'stochasticity' of quantum mechanics adds only randomness to decision-making processes, not freedom. In contrast, the theory of computation implies that even when our decisions arise from a completely deterministic decision-making process, the outcomes of that process can be intrinsically unpredictable, even to—especially to—ourselves. I argue that this intrinsic computational unpredictability of the decision making process is what gives rise to our impression that we have free will."

12. The interdisciplinary nature of the Milan conference meant that it was aimed at a wide audience, at bringing people together from many different backgrounds, so it was important not to limit the discussion only to the ontological status of determinism and free will. On the contrary, we tried also to investigate the connections that these concepts have with a variety of different human experiences, albeit in the limited space of one afternoon.

Of course, these interconnections would have required many different speakers to ensure that every point of view could be expressed. Clearly, it would have been a vain hope even just trying to achieve anything like completeness here, when we consider all the disciplines involved, i.e., biology, law, economy, neuroscience, history, theology, etc.

However, the rather relational nature of the theological point of view, as well as the happy contingency of the 500th anniversary of the publication of the Theses of Luther (1517–2017), suggested to us to complete the trio of speakers with the theologian Piero Coda.

In fact, the contribution of the Christian theological tradition to the formation of the very concept of freedom proper to modern Western civilization cannot be ignored. In his essay Coda illustrates how the concept of freedom emerges in the theological debate, in interaction with the concepts of grace and relation. This evolution is also vividly depicted through a comparison with the classical Greek concept of fate, which envisaged, for humans and even for gods, a much more limited level of freedom in their actions, and therefore in their responsibilities, than what was instead predicated by the Christian doctrine in the first centuries of our era.

Of course, the languages and conceptual horizons of the three speakers were profoundly disparate, as the reader will easily realize in the following essays. However, these differences did not prevent an attempt at dialogue and interaction, an attempt that actually raises many more questions, rather than bringing many answers. But it seemed to us that this should in fact be one of the main objectives of an event conceived at the outset as multidisciplinary rather than specialized, and aimed at the general public.

13. Some considerations that may perhaps also generate theological reflections are as follows. A universe in which free will exists (not illusory, as described by Seth Lloyd, but fundamental) is necessarily a non-deterministic, a-causal universe in which at least some

events (at least some in the mind of the person who makes the choice) happen, radically, by chance (otherwise free will, namely a choice free from the influences of past events, would not exist). On the contrary, a rigidly deterministic universe is guided by the most absolute and indisputable principle of causality. It is basically a universe whose entire history is given from the beginning, a universe in which "past and future would be equally present in the eyes of an intelligence equipped with superior analytic capabilities," as Laplace wrote. It would therefore be an eternal universe, whose events are already all given and set from the beginning, in which time, in the sense of "becoming", is absent. This dualism has an interesting resemblance to the biblical myth of Adam and Eve's "fall" in the earthly paradise. Before choosing whether or not to eat the forbidden fruit, Adam and Eve lived in heavenly conditions, where all worries, pains, anguishes, and death were absent. A state of eternal idyllic present, a timeless condition of eternity. The choice of whether or not to eat the forbidden fruit, the exercise of free choice, destroys paradise and with it eternity. Human history, with all the joys and all the pains of the human condition, began at that moment, with that exercise of free choice. The biblical story, from this point of view, seems to link the beginning of human history to an act of free choice.

14. Finally, it is worth reporting Vervoort's idea about the origin of probabilistic distributions as a product of underlying (micro) deterministic laws:

"It is straightforward to observe that any variable that is the effect of many hidden (and independent) causes, or in other words that is a function of many hidden and independent causes, is normally distributed, independently of the distribution of the causes (the latter may even have extremely simple, fully non-random-looking distributions). In other words, '(normal) probability' is interpreted in terms of the occurrence of causes, more precisely, in terms of the massive averaging effect of a large number of causes. One is tempted to say that anything that has many causes will look probabilistic.

If the majority of probabilistic properties of the world, namely those described by the normal distribution, can be understood as emerging from individual deterministic processes, it is tempting to conjecture that they all can.

There is a literally infinite number of probabilistic systems, from such diverse areas as fluid mechanics, diffusion, ballistics, error theory, population dynamics, population statistics, games of chance, quantum mechanics, information processing, and every field of engineering. All these profoundly different systems show the same frequency stabilization, the same need to converge towards well-defined ratios. They all obey the same simple laws of probability theory. The only possibility I can imagine to explain this 'necessity', shared by all these systems, is that they share the necessity of laws governing the evolution of their individual constituents, i.e., the necessity of determinism."

I thus conclude my introduction, which has perhaps become too long. However, with the presentation of several ideas, I hope it does what every introduction should do, namely, increase the reader's curiosity for the issues discussed by the authors, and encourage further reading, in this or other volumes.

Bibliography

J.S. Bell, *Speakable and Unspeakable in Quantum Mechanics* (Cambridge University Press, Cambridge, 1987)

J. Conway, S. Kochen, The free will theorem. Found. Phys. **36**, 1441 (2006)

A. Einstein, *Opere Scelte* (Bollati Boringhieri, Torino, 1988)

A. Einstein, B. Podolsky, N. Rosen, Can quantum mechanical description of physical reality be considered complete? Phys. Rev. **47**, 777 (1935)

G. 't Hooft, *The Free Will Postulate in Quantum Mechanics* (2007). arXiv:quant-ph/0701097

G. 't Hooft, *The Cellular Automaton Interpretation of Quantum Mechanics* (Springer Open, Berlin, 2016)

S. Hossenfelder, Free will is dead, let's bury it (2016). http://backreaction.blogspot.nl/2016/01/free-will-is-dead-lets-bury-it.html

S. Lloyd, A Turing test for free will. Phil. Trans. Roy. Soc. **A28**, 3597 (2012)

A. Pais, *Subtle is the Lord: The Science and the Life of Albert Einstein* (Oxford University Press, Oxford, 1982)

L. Russo, *The Forgotten Revolution* (Springer, Berlin, 2004)

E. Schroedinger, Die gegenwaertige Situation in der Quantenmechanik, Die Naturwissenschaften **23**, 807–812, 823–828, 844–849 (1935)

E. Severino, *Legge e Caso* (Adelphi Edizioni, Milano, 1979)

E. Severino, *The Essence of Nihilism* (Verso Books, London, 2016)

L. Vervoort, Does chance hide necessity? Ph.D. thesis, 2014. arXiv:1403.0145

M. Zych, F. Costa, I. Pikovski, C. Brukner, Bell's theorem for temporal order (2017). arXiv:1708.00248

Free Will in the Theory of Everything

Abstract From what is known today about the elementary particles of matter and the forces that control their behaviour, it may be observed that a host of obstacles to our further understanding remain to be overcome. Most researchers conclude that drastically new concepts must be investigated, new starting points are needed, older structures and theories, in spite of their successes, will have to be overthrown, and new, superintelligent questions will have to be asked and investigated. In short, they say that we shall need new physics. Here, we argue in a different manner. Today, no prototype, or toy model, of any so-called theory of everything exists, because the demands required of such a theory appear to be conflicting. The demands that we propose include locality, special and general relativity, together with a fundamental finiteness not only of the forces and amplitudes, but also of nature's set

Presented at the Workshop on 'Determinism and Free Will', Milan, May 13, 2017.

of dynamical variables. We claim that the two ingredients we have today, quantum field theory and general relativity, do indeed go a long way towards satisfying such elementary requirements. Putting everything together in a grand synthesis is like solving a gigantic puzzle. We argue that we need the correct analytical tools to solve this puzzle. Finally, it seems obvious that this solution will leave room neither for 'divine intervention', nor for 'free will', an observation that, all by itself, can be used as a clue. We claim that this reflects on our understanding of the deeper logic underlying quantum mechanics.

Theories of Everything

What is a 'theory of everything'? When physicists use this term, we begin by emphasising that this should not be taken in a literal sense. It would be preposterous for any domain of science to claim that it can lead to formalisms that explain 'everything'. When we use this phrase, we have a deductive chain of exposition in mind, implying that there are 'fundamental' laws describing space, time, matter, forces, and dynamics at the tiniest conceivable distance scale. Using advanced mathematics, these laws prescribe how elementary particles behave, how they exchange energy, momentum, and charges, and how they bind together to form larger structures, such as atoms, molecules, solids, liquids, and gases. The laws have the potential to explain the basic features of nuclear physics, astrophysics, cosmology, and materials science. With statistical methods they explain the basis of thermodynamics and more. Further logical chains of reasoning connect this knowledge to chemistry, the life sciences, and so on. Somewhat irreverently, some might try to suggest that a 'theory of everything' lies at the basis of most of the other sciences, while of course this is not the case. We must avoid giving the impression that the other sciences would be thought of as less 'fundamental'. In practice, a theory of everything would not much affect the rest of science, simply because each of the elements of such a deductive chain would be far too complex and far too poorly understood to be of any practical value. The theory applies to 'everything' only in a formal sense.

What do physicists imagine a 'theory of everything' to look like? Should it be a 'grand unified theory' of all particles and forces? If so, we are still a long way off, since the relevant distance scale at which fundamental modifications to our present theoretical views are expected to be needed is the so-called Planck length, some 10^{-33} cm, which is more than a billion times a billion times smaller than anything that can be studied directly in laboratory experiments. Is it 'quantised gravity'? Deep and fundamental problems arise when we try to apply the principles of quantum mechanics to the gravitational force. Forces and quantum mechanical amplitudes tend to infinity, and the remedies for that, as proposed so far, still seem to be very primitive. Since they lead us out of the perturbative regime, calculations are imprecise, and accurate definitions explaining what we are talking about are still lacking.

Is it 'superstring theory'? The problem here is that this theory hinges largely on 'conjectures'. Typically, it is not understood how most of these conjectures should be proven, and many researchers are more interested in producing new conjectures rather than proving old ones, as this seems to be well nigh impossible. When trying to do so, one discovers that the logical basis of such theories is still quite weak. One often hears the argument that, although we do not quite understand the theory 'yet', the theory is so smart, that it understands how it works itself. Or again, its mathematics is so beautiful and coherent that it 'must be true'. In the view of the present author, such arguments are more dubious than often realised, if the history of science is anything to go by.

Finally, many researchers are tempted to part from the established paths to try 'completely new and different' starting points. In many cases, these are not based on sound reasoning and healthy philosophies, and the chances for success appear to be minimal. The reader may not realize at first glance, but the present paper is a plea for rigorous reasoning and carefully keeping established scientific results in mind.

Is humanity smart enough to fathom the complexities of the laws of nature? If history can serve as a clue, the answer is: perhaps. We are equipped with brains that have evolved a little bit since we descended from the apes, hardly more than a million years ago, and we have managed to unravel some of nature's secrets way beyond what is needed to build our houses, hunt for food, fight off our enemies, and copulate. In terms of cosmic time units, a million years is not much, and our brains may or may not have had enough opportunity to evolve to a state where they can carry out this particular new task. However, we may just about manage to figure things out, making numerous mistakes on our way. And the nice thing about science is that mistakes can be corrected, so we do stand a reasonable chance.

Today's attempts at formulating 'theories of everything' must look extremely clumsy in the eyes of beings whose brains have had more time, say another few million years, to evolve further. The present author is convinced that many of the starting points researchers have investigated up to now are totally inappropriate, but that cannot be helped. We are just baboons who have only barely arrived on the scene of science. Using my own limited brain power, I am proposing a somewhat different starting point.

The following two sections, the main body of my lecture, may look peculiar, contemplating the laws of nature from an unconventional vantage point. We argue that the fundamental laws of nature appear to be chosen in an extremely efficient way. The only thing that may seem *not* to agree with our philosophy is quantum mechanics. On the other hand, quantum mechanics does also appear to be an extremely efficient theory. Without quantum mechanics, we would not have been able to construct meaningful theories for atoms and sub-atomic particles.

The good thing about quantum mechanics is the simple fact that many of nature's variables that used to take continuously varying values in classical physics now turn out to be quantised. In Sect. 5 we summarise observations concerning the mathematical coherence of quantum mechanics. One could ascribe the very special logical structure of quantum mechanics to the inherent discreteness of the physical variables it describes. Now this special form of logic also seems to force us to abandon the notion of definiteness of observables, as if nothing can be absolutely certain in a quantum system. But looking deeper into the mathematical structure of the theory, one can question such conclusions. The author has somewhat different views on quantum mechanics, which we briefly explain in Sect. 6.

Our conclusion will be that our world may well be superdeterministic,[1] so that, in a formal sense, free will and divine intervention are both outlawed. However, we emphasise that, in daily life, nobody will suffer from the

---

[1]Here, superdeterminism is not intended to mean *pre*-determinism, the idea that the future may be fixed prior to the action of physical law, but rather ordinary determinism that also has to apply to any observer who chooses what to observe.

consequences of such an observation; it pertains to the deeper fundamental nature of physical laws.

God's Assignment

Imagine that you are God.[2] Your assignment is: *run a universe*. Your universe may look like a big aquarium, containing things like stars and planets, plants, animals, humans, and more elementary objects such as atoms and subatomic particles. To make it all work, you will want billions or more of all of these. You may steer all these objects in any way you like, and you want interesting things to happen. What should you do?

You would have a problem. To tell every individual object in your universe what to do will require a massive amount of administration. Suppose you want to be efficient, isn't there an easier way? The answer is yes. You declare that there are rules. Every object, every particle this object is made of, moves around as directed by *laws of nature*. Now, there are only two things left to be done: design laws of nature, and obtain a powerful computer to help you implement the laws of nature. Let us assume that you have such a powerful computer. Then the question is: how do you choose the laws of nature?

Stars, planets, and people are quite complex, so you do not want the rules to be too simple, since then nothing of interest will happen in your universe. Computer scientists would have ideas about designing rules, a software routine, a program, telling you how your universe evolves, depending on the laws you feed it with, but to make your universe sufficiently realistic, their programs will tend to become lengthy, complex, and ugly. You want to be more demanding.

---

[2]This is only meant metaphorically; this author, fortunately, is not religious.

So, being God, you have a second great idea. Before formulating your laws of nature, you decide about a couple of *demands* that you impose upon your laws of nature. Tell your computer scientists and mathematicians that they must give you the simplest laws of nature that comply with your demands. While listening to what your computer scientists and mathematicians tell you about the viability of your demands, you copy the rules they formulate. You impose the rules, and you press the button.

In this paper, it will be argued that very simple demands can be imposed, and that at least some of these demands already lead to a structure that may well resemble our universe. The construction that will eventually emerge will be called the 'theory of everything'. It describes everything that happens in this universe.

Now, it will appear at first sight that the first demand suggested here will not be obeyed by the actual universe. But these are only appearances. Remember that our brains were not designed for this, so keep your prejudices in check for the moment. We claim to be able to make three observations:

- The set of demands that we will formulate now are nearly inevitable and non-negotiable.
- Even though the demands are simple, the mathematical structure of the rules, or laws of physics, will turn out to be remarkably complex, possibly too complex for simple humans to grasp.
- As far as we do understand them, the resulting rules do resemble the laws of nature governing our actual universe. In fact, it may well be that they lead *exactly* to our universe.

This is my projected path towards a 'theory of everything'.

Demands and Rules

> **Demand #1**: Our rules must be unambiguous. At every instant, the rules lead to a single, unambiguous prescription of what will happen next.

Here, most physicists will already object: *What about quantum mechanics?* Our favoured theory for the sub-atomic, atomic, and molecular interactions dictates that these respond according to chance. The probabilities are dictated precisely by the theory, but there is no single, unambiguous response.

I would make three points here. The first is that this would be a natural demand for our God. As soon as he admits ambiguities in the prescribed motion, he will be thrown back to the position where gigantic amounts of administration are needed: what will be the 'actual' events when particles collide? Or alternatively, God would have to do the administration for infinitely many universes all at once. This would be extremely inefficient, and when you think of it, quite unnecessary. God would much prefer a single outcome for any of his calculations. This, by the way, would also entail that his computer be a *classical* computer, not a quantum computer (Zuse 1969; Fredkin et al. 2003; Feinstein 2017).

The second point is this. Look at the universe we live in. The ambiguities we have are in the theoretical predictions about what happens when particles collide. What *actually* happens is that every particle involved chooses exactly one path. So God's administrator must be using a rule for making up his mind when subatomic particles collide.

The third point is that there are ways around this problem. Mathematically, it is quite conceivable that a theory exists that underlies quantum mechanics (Hooft 2016).

This theory will only allow single, unambiguous outcomes. The only problem is that, at present, we do not know how to calculate these outcomes. I am aware of the large numbers of baboons around me whose brains have arrived at different conclusions: they *proved* that hidden variables do not exist. But the theorems applied in these proofs contain small print. It is not taken into account that the particles and all other objects in our aquarium will tend to be strongly *correlated*. They howl at me that this is 'super-determinism', and would lead to 'conspiracy'. Yet I see no objections against super-determinism, while 'conspiracy' is an ill-defined concept, which only exists in the eyes of the beholder. A few more observations on this topic are made in Sect. 6.

> **Demand #2**: We must have *causality*: every event must have a cause, and these causes must all lie in the past, not in the future.

A demand of this sort is mandatory. What it really means is that, when our God looks up his rules to figure out what is supposed to happen next, he should never be confronted with a circular situation, or in other words, he should always know in which *order* the rules must be applied. Whatever that order is can be used to define *time*. So now we can distinguish future from past. Only the past events are relevant for what happens next, and whatever they dictate, will only affect the future. This principle has been instrumental in helping us understand quantum field theories, for instance.

> **Demand #3**: *Efficiency*: Not all events in the past, but only a few of them, should dictate an event in the present.

This suggests that there is a power limitation in God's laptop. We cannot have a situation where the complete past history of every particle is necessary to determine the behaviour of a given particle in the future. If, for computing the behaviour of one particle, we only need the data concerning a few particles in its immediate environment, then the calculation will go through a lot more quickly. We are simply asking for a maximum of efficiency; there are still a lot of calculations to do.

This demand will now also lead to our first rule, or law of physics:

> **Rule #1**: *Locality*. All the configurations one needs to know to determine the behaviour of an object at a given spot must lie in its vicinity.

This means that we will have to define *distances*. Only points at very small distances from a given point are relevant to what happens there. What 'vicinity' really means still has to be defined. In our universe it is defined by stating that space is three-dimensional, with a Euclidean definition of distance. Details will be left for later.

Now we still have to decide how things interact, but before that, we have to decide how things can move. This is a delicate subject. In our universe, things that can stand still will be allowed to move along straight lines, with, at first sight, any speed. This makes things difficult for God's computer programmer. A programmer will find it easy to define objects that move with a predefined speed in a predefined direction, but what are the rules if something may move with *any* speed in *any* direction? A deceptively simple-looking answer comes in the form of a new law of nature:

> **Rule #2**: *Velocity*. Any object for which it has been decided how it behaves when at rest, will behave very similarly when it moves along a straight line in any direction, with any constant speed (within limits, see the next rule).

The rules governing its behaviour at this velocity must be derivable in a simple way from the rules governing its behaviour when it stands still. Think of someone sitting in a train, playing chess. How this person feels, how he moves his arm while moving a pawn, as well as the rules for chess, all are the same when the train moves as they are when the train stands still.

This is an important rule, since it builds an enormous amount of structure and complexity into our universe. The relative positions of things can now change with time, but things can also collide, they can have moving parts, etc. At first sight, the price we pay for this added complexity may seem to be mild, since we get moving things for free, if we know how they function when standing still.

But there is a problem. Should we accept *all* speeds, or should there be a speed limit? If we don't impose a speed limit, we run into trouble. The trouble is with locality. If things move infinitely fast, they can be simultaneously here and far away. In practice, this also means that there will be trouble with our demand of locality and/or efficiency. Many of our particles will move so fast that our computer needs infinite processing speed. This we should not allow, so we impose:

Rule #3: There is a *speed limit*. Call it c, the speed of light.

This fits well with locality: the only neighbours that interact with a particle of matter are the ones that can be reached by a light signal within a limited time step.

We know that there is a speed limit in our universe: the speed of light. Thus, we saved efficiency and locality, but now there is a new problem. The person in the train moves his arm. The train may be moving more slowly than the speed limit, but what about the arm? Well, God's mathematicians tell God that this problem can be solved, but a number of amendments must be made to rules #2 and #3. First, let us slow the arm down:

Rule #3a: Things that go faster will experience time going more slowly.

This slows down the arm, but it is not quite enough. We also need:

Rule #3b: Things in motion will also contract in the forward direction.

This makes the arm shorter, but again it does not help quite enough. A more drastic measure:

Rule #3c: Inside moving things, clocks will no longer go synchronously.

In combination with rules #3a and #3b, this works: a person inside the train may stick his arm out, but as seen from outside the train, the arm reaches the pawn at the moment that the body has nearly overtaken the arm. Mathematicians tell us that we need all three of these amendments to rule #3, and now the logic works out fine; the arm will not exceed the speed limit.

Physicists have learned about this rule, with its amendments, along somewhat different lines of reasoning, but the result is the same: Einstein's special relativity. As we see here, Einstein's relativity theory could have been deduced from purely logical arguments, if our brains had been hundreds of times smarter. In our world, it was arrived at by Hendrik Antoon Lorentz, by studying the laws of electromagnetism. We see now why we could never have understood our universe if there hadn't been special relativity.

We still haven't tried to determine *how* things behave, even if everything stands still. This is because there is another problem. Think of an object, such as a lump of sugar. It is extended in space. Because of locality, one side of our lump of sugar should behave independently from the other side. Should it not be possible to break the lump of sugar in half? And the pieces we then get, should we not be able to break these in half again? And so on? Can we break these pieces in half forever?

This question was already raised by the Greek philosophers Democritus, Leucippus, and Epicurus around 400 BC. In a stroke of genius, they stumbled upon the right answer: *No, this series of divisions will stop*. There will be a smallest quantity of sugar. They called these smallest quantities 'atoms'. So it was these Greeks who first tried the concept of *quantisation*. They quantised matter, purely by using their brains. Our God is forced to assume something like this as well, since, if things could be divided into pieces ad infinitum, this would imply infinite complexity, which needs to be avoided.

In the name of our efficiency requirement, we must quantise as well. All objects in this universe can be broken into smallest units. The 'atoms of sugar' are now known as 'molecules', but that's just a detail. Molecules were found to be composed of smaller things, and these are now called 'atoms', which can be divided further: the smallest possible objects are now called elementary particles. Note that, in modern theories of elementary particles, these particles are considered to occupy single points in space. A point cannot be divided in two. However, when a particle such as an electron, emits another particle, for instance a photon, then in particle physics we say that the photon is created at that spot; it does not hide inside an electron. In conclusion, it turns out that we need:

Rule #4: Matter is *quantised*. The smallest quanta of matter are the elementary particles.

But for God's mathematicians, these quanta of matter have caused considerable trouble. The quanta will probably carry mass, energy, and momentum, but even if they are point-like, we sometimes do need a property replacing the notion of size. It was found how to do this: for all particles with finite amounts of momentum, there is a natural smallest size limit. The math needed is called quantum mechanics.

Quantum mechanics works, but it is complicated. Yet, up to this point, today's physicists and mathematicians have discovered how to combine these rules in a working configuration. In particular, adding special relativity, rule #3, took us nearly 50 years, so it wasn't easy. The result was called 'quantum field theory'. There is one problem with quantum field theory: it is not known where the *forces* come from; this leaves us with lots of freedom, as it is not known how God made his decisions here.

Thus, we have to introduce one more concept: forces. In our universe, it must be possible to *change* the velocities of objects, and decide about a rule for this, of the following type:

> **Rule #5**: *Forces.* If it is known how an object behaves while moving with constant speed on a straight line, it should be possible to deduce how it behaves while moving with a varying speed on a curved line.

The primary force that can be deduced in this way is gravity. In our world, we know that other forces exist, but these may be due to secondary effects resulting from complex behaviour at ultrashort distances. Again, there will be a price to pay: the best way to add the notion of curved lines in the logic of our rules is to have curvature in the fabric of space-time itself. One may then create the situation that the fundamental differences between straight lines and curved lines disappear; on curved spaces, straight lines do not exist.

There is also another advantage: curved universes have no fixed size, they can expand. This means that our universe may begin by being very tiny and very simple, and grow all by itself, just by the action of our rules. In our universe, this situation occurs. The theory describing these aspects is Einstein's theory of general relativity.

This leads us to one more rule:

Rule #6: God must tell his computer what the *initial state* is.

Again, efficiency and simplicity will demand that the simplest possible choice is made here. This is an example of Occam's rule. Perhaps the simplest possible initial state is a single particle inside an infinitesimally small universe.

Final step:

Rule #7: Combine all these rules into one computer program to calculate how this universe evolves.

So we're done. God's work is finished. Just push the button. However, we have reached a level where our monkey brains are at a loss. Rules #5 to 7 have proven to be too difficult for us. The theory of general relativity manages to take Rule #5 into account, but unfortunately does not handle quantum mechanics, rule #4, correctly.

Why is this so difficult? Quite possibly, more rules will have to be invented to reach a coherent evolution law, but up to now, we have been confronted with this question. Can we implement all the rules given above into a single, working scheme? What comes out will be of secondary importance. Perhaps a framework will be found with many possibilities (a 'multiverse'). In that case, more rules will have to be invented to single out one preferred choice.

So far, it seems that the requirements we mention above have all been taken into consideration in the laws of physics of our universe. This is why this author suspects that the given rules make a lot of sense.

We note that the actual laws of physics known to hold in our universe are quite close to what we have constructed purely by mental considerations. Of course, the author admits that this will be attributed to hindsight, but we claim that a super intelligent entity could perhaps have 'guessed' nature's laws of physics from such first principles. This would be important to know, since this would encourage us to use similar guesses to figure out how the remaining physical laws, not yet known to us today, might also be guessed.

The idea of underlying laws that are completely mechanical, while what we currently know as quantum mechanics should be an emergent feature of the universe, has been suggested several times (Zuse 1969; Fredkin et al. 2003; Feinstein 2017), but there are deep problems with it, which will be addressed now.

Free Will

Note what has motivated the demands formulated in Sect. 3: unambiguity, simplicity, efficiency, and finiteness. In particular this last demand, finiteness, is not (yet?) completely implemented in the known laws of nature today. There are various things that can go out of control due to infinities. In quantum field theories, we managed to keep one kind of infinities under control, the infinities in the quantum amplitudes and all physical effects associated with those. This means that the effects of forces in the theory stay finite and computable.

This is important. However, we always need to restrict ourselves to approximations, in this case, perturbation expansion techniques. An infinity that we left aside because, on the face of things, it did no harm, is the infinity of all the relevant dynamical physical variables. For us this seems to cause no problems, as long as our integrals converge, but for a 'God' who wishes to stay in control of everything, this is not an option: the total number of independent variables must be finite. His laptop must be able to compute *exactly* what happens in a finite stretch of time. Here, our arguments seem to favour a universe that is spatially compact rather than unbounded.

The reader might accuse me of an ill-motivated religious standpoint, but we do note that the rules we arrived at by using this standpoint are remarkably effective in generating laws of nature that are known to work quite well.

Then, the reader might point out that no classical laptop at all can compute quantum mechanical amplitudes with infinite precision, and insist that a *quantum* laptop would be needed. This however, might be the result of an elementary incompleteness in our present understanding of quantum mechanics, as we argued some time ago ('t Hooft 2016). There is every reason to suspect that a novel theory underlying quantum mechanics will be required. To satisfy our demand of unambiguity, all phenomena must be entirely computable, not left to chance.

If this is right, the laws of nature we arrive at leave no room for two things:

- divine intervention, and
- free will.

We claim that there would be very little justification for the existence of either. If we allowed for divine intervention, for instance in all quantum mechanical phenomena, our theory would leave such a gigantic amount of arbitrariness in its prescriptions that all the laws of physics would seem to be there for no particular reason. We would find ourselves back at square one. As for free will, the argument is very similar. If quantum mechanics left room for free will, there would be far too much room for it. There is every reason to suspect that today's voids in the theory of quantum mechanics will be filled by additional laws.

Most importantly, quantum mechanics itself can be used to show us how the voids might be filled in. It is not hard to imagine versions of our dynamical theories where quantum mechanics as it appears today can be aptly described in classical terminology, but we need these missing laws.

It is important to note that quantum mechanics accurately predicts the statistics we observe when experiments are repeated many times. If there are additional laws that decide about individual events, these laws must reproduce the statistics as predicted by quantum mechanics alone. This implies that the question whether the additional laws exist or not will not be decidable experimentally. Physicists who are content with a theory that never gives better answers than statistical ones will categorically reject speculations concerning hidden variables, but religious people who assume that our universe is reigned over by some God should require quantum mechanics to be supplemented with theories of evolving hidden degrees of freedom, in such a way that all events that take place can be attributed to something that has happened nearby in the past.

Whether devine intervention takes place or not, and whether our actions are controlled by 'free will' or not, will never be decidable in practice. It is thus suggested here that, where we succeeded in guessing the reasons for many of nature's laws, we may as well assume that the remaining laws, to be discovered in the near or distant future, will also be found to agree with similar fundamental demands. Thus, our suspicion of the absence of free will can be used to guess how to take the next step in our science.

Quantum Mechanics

Today's scientists have not yet reached that point. All dynamical laws in the world of (sub)atomic particles have been found to be controlled by quantum mechanics. Quantum mechanics appears to add a sense of 'uncertainty' to all dynamical variables describing these particles: the positions and momenta of particles cannot be sharply defined at the same time, the components of the spin vector contain a similar notion of uncertainty, and sometimes the creation of a particle can be confused with the annihilation of an antiparticle—and so on. This is actually not a shortcoming of the theory, because in spite of these apparent uncertainties, the statistical properties of the elementary particles can be determined very precisely. So what is going on?

The rules according to which quantum mechanics works are precisely formulated in what is sometimes called the Copenhagen interpretation. This is not the place to explain what the Copenhagen rules are, but they can be summarised by stating that the behaviour of a particle can be described as completely as if there were no 'uncertainties' at all. Instead, we have variables that do not commute:

$$x \cdot p - p \cdot x = [x, p] = i\hbar. \tag{1}$$

In practice, what this means is that, when a system is described quantum mechanically, we apply a number system that is more general than in classical physics, but just as applicable. In fact, this number system is more useful than the old, commuting numbers, when it is used to refer to quantities that are quantised, i.e., that only come in integer multiples of fixed packages.

Thus, imagine a system that allows its dynamical variables only to occur in distinct states, typically indicated by integer numbers, $|1\rangle, |2\rangle, |3\rangle$ .... The non-commuting numbers that we use can then be called *observables* when they simply describe the state the system is in, like indicating the value that a particular integer has. When they are applied to replace a state by another state, they are called *operators*; for instance, $a|n\rangle = |n-1\rangle$. The manipulations we use to handle these numbers act the same way regardless of whether we are dealing with observables or operators. This makes quantum mechanics extremely flexible, but it sometimes obscures the situation when we are unable to distinguish observables from operators.

A *quantum transformation* replaces observables by operators, or more often, mixes the two types completely. One ends up having to use wave functions to describe the states a system can be in. The beauty of this formalism is that such numbers, the non-commuting numbers, regain continuity even if the original system was discrete, and thereby allow us to use the machinery of advanced mathematics. This leads to such powerful results that few physicists are ready to return to the original system of discrete physical states describing 'reality'. After most quantum transformations, reality is replaced by the more abstract notion of a wave function. This notion only appears to be abstract if we ask 'What is going on here?', but in practice serves us very well if we only ask 'What will the result of this experiment be?'

In fact, according to the Copenhagen interpretation, questions such as 'What is going on here?' are ill-posed questions, as they cannot be answered by doing experiments. In practice, therefore, we usually refrain from asking such questions. All that matters is the reproduction of the answers given by experiments.

Nevertheless, our question 'What is gong on here?' is not ill-posed. We can always attempt to find answers of principle: we do not know what is going on here, but we can imagine very precisely what it could be. *Something* is going on, and the assumption that there is something going on that might explain what is happening next, even if we cannot be certain what it was, may be used as an important constraint in constructing theories. A typical example is the Standard Model of the sub-atomic particles. This model was established partly by doing experiments with elementary particles, but also by imagining how these particles should behave. It makes sense to use as an assumption: every particle behaves in a completely deterministic way, even though its behaviour cannot be completely determined by any known observation technique. The assumption that particles behave in such a way that a completely deterministic theory is responsible is not a crazy assumption, but it requires guesswork. In principle, such guesses could help us to guess correctly what the next stage for the Standard Model might be.

Bell's Theorem

The reasoning summarised above, which I explained more elaborately in ('t Hooft 2016), may seem to be logical, yet it is nearly universally rejected by researchers in quantum mechanics. The reason for that is that there seems to exist a rigorous proof of the contrary statement: *experiments can be carried out, for which standard quantum theory provides very firm predictions concerning their outcomes, predictions that have indeed been confirmed by experiment, while they do not allow for any 'ontological' description at all.* The question 'What is going on here?' cannot be answered at all without running into apparent contradictions.

For a complete description of J. S. Bell's Gedanken experiment, we refer to the literature, and references therein (Bell 1964; Clauser et al. 1969). Here, we summarise. Bell's starting point is that experimenters can put small particles in any quantum state they like, and this appears to be true in most cases. In particular, we can put a pair of particles in an *entangled* quantum state. Photons, for example, are described not only by a plane wave that determines in which direction a photon goes, but also by their polarisation state. Photons can be linearly polarised or circularly polarised, but there are always exactly two possibilities for the polarisation: vertically or horizontally, or alternatively, left or right circularly polarised.

An atom can be put in an excited state in such a way that it emits two photons, which, together, form only one possible quantum state: if one photon is found to be vertically polarised, the other will necessarily be vertically polarised as well, and if one photon is circularly polarised to the left, the other is polarised to the left as well. In this case, the two photons form an entangled state. Together, however, this is only a single, allowed quantum state that the pair of photons can be in.

Far from the decaying atom, Bell now imagines two detectors, called Alice and Bob, each monitoring one of the photons. They both work with linear polarisation filters, checking the polarisation of the photon that they found. They do a series of experiments, and afterwards compare their results. They do not disclose in advance how they will rotate their polarisation filters. Now, whenever the two polarisation filters happen to be aligned, it turns out that they both measure the same polarisation of their photons. When the two polarisation filters form an angle of 45°, they find the two photons to be totally uncorrelated. But when the relative angle is 22.5° or 67.5°, they find a relatively high correlation of the two polarisations. In classical physics, no simple model can be constructed that reproduces this kind of correlation pattern.

The only way to describe a conceivable model of 'what really happens', is to admit that the two photons emitted by this atom know in advance what Bob's and Alice's settings will be, or that, when doing the experiment, Bob and/or Alice know something about the photon or about the other observer. Phrased more precisely, the model asserts that the photon's polarisation is *correlated*[3] with the filter settings later to be chosen by Alice and Bob. We can compute what kind of correlation is needed. One finds that the correlation is a pure *three-body correlation*: if we average over all possible polarisations of the photon pair, we find that Alice's and Bob's settings are uncorrelated. If we average over all possible settings Alice can choose, then Bob's settings and the polarisation of the photons are again uncorrelated, and vice versa.

But this three-body correlation is said to be impossible. How can the photons know, in advance, what Bob and Alice will do? In deriving his inequalities, Bell, and later Clauser, Horne, Shimony, and Holt (Clauser et al. 1969), assumed that the polarisation state of the entangled photons was independent of the settings chosen by Alice and Bob. This assumption has been discussed many times in the literature. It was subsequently concluded that it is

---

[3]This can also be rephrased as follows: the assumption that, in the initial state, the experimenter can always produce any entangled state of photons as he or she pleases is not true. There will be strong and uncontrollable—entangled—correlations with other atoms in the system.

inevitable, but, as we shall argue, there can be correlations, and they must be strong. This invalidates the inequalities derived by Bell and CHSH. Bell's theorem is violated because these inequalities are violated. What remains to be done is to explain how this could have happened, because the required correlations seem to run against common sense; they appear to contradict the notion that, after the photons have been emitted, both Alice and Bob have the *free will* to choose any setting they like. But have they?

It is easy to say that they have not. If we adhere to a deterministic model, it is clear that the polarization of the photon, as well as the settings chosen by Alice and Bob, have been determined by the initial state of the universe, together with deterministic equations of motion. But this is not the complete answer to our problem. How do we make a model for these photons?

Apparently, what quantum mechanics dictates is a strong 3-body correlation. The three points in space and time that are correlated may well all be spatially separated from one another. This means that no signal can have been transmitted from one to the other, but this is not a problem. It is well known in the quantum theory of sub-atomic particles that correlations need not vanish outside the light cone.[4] The real problem here is that Alice's and Bob's settings are classical, and the quantized atom was there first. What kind of model can bring about such

---

[4]In contrast, the *commutator* of two operators defined on a pair of space-time points does vanish outside the light cone. The commutator can be seen to monitor causal influences of one operator on the value of the other, and so one can prove that a non-vanishing commutator will enable experimenters to send signals to one another, while the *correlation function* only points towards a common past of the pair of space-time points. Bell was aware of the use of commutators to define causality ('no Bell telephone'), but he needed something stronger, which we now refute.

strong correlations, even if they are 3-point correlations, when two of the variables considered are classical?

If this is the way to look at the problem raised by Bell's theorem, we can limit ourselves to a more elementary question. Consider just a single, polarised photon. It may have been emitted by some quasar, billions of years ago. An observer detects it after it has passed a polarisation filter. The photon either passes or it does not. In both cases, the 'true polarisation state' of the photon was either in line with the observer's filter, or orthogonal to it, but not in any other direction. It seems as if the quasar, billions of years ago, already knew that these were the two polarisation directions the photon had to choose from. This will be a strange aspect of any model that we might want to apply.

And now for what this author believes to be the correct answer, both for the single photon problem and the Bell experiment. Our theory is that there does exist a true, ontological state, for all atoms and all photons to be in. All ontological states form an orthonormal set, the elements of an ontological basis. The universe started out in such a state, and its evolution law is such that, *at all times in the future, the universe will still be in an ontological state.* Regardless of which ontological initial state we start from, the state in the future will be an ontological one as well, that is, *not a quantum superposition of different ontological states.* What we have here, is a conservation law, the conservation of ontology. It selects out which quantum superpositions can be allowed and which not, just because, according to our model, the evolution law is ontological.

An ontological photon can be polarised in any sort of way, but it cannot evolve into any superposition of ontological states, and this law is universal, it holds for all states the universe can be in. The outcomes of both Alice's and Bob's measurements are ontological, so this ensures that the photons they look at, including ones that can have travelled billions of years, have been ontological at all times. What is not widely known is that this conservation rule is also respected by the Schrödinger equation, so that no modification of quantum mechanics is necessary.

The effect of this law is so strong that it looks like 'conspiracy', but this law is not more conspiring than the law of conservation of angular momentum. The correlation function needed in a simple model for Bell's Gedanken experiment was calculated in ('t Hooft 2016). Our argument is similar to several raised earlier, such as (Vervoort 2013).

Conclusion

The author agrees with Bell's and CHSH's inequalities, as well as their conclusions, given their assumptions. We do not agree with the assumptions, however. The main assumption is that Alice and Bob choose what to measure, and that this should not be correlated with the ontological state of the entangled particles emitted by the source. However, when either Alice or Bob change their minds ever so slightly in choosing their settings, they decide to look for photons in different ontological states. The free will they do have only refers to the ontological state that they want to measure; this they can draw from the chaotic nature of the classical underlying theory. They do not have the free will, the option, to decide to measure a photon that is not ontological. What will happen instead is that, if they change their minds, the universe will go to a different ontological state than before, which includes a modification of the state it was in billions of years ago.[5] Only minute changes were necessary, but these

---

[5]The new ontological state cannot have overlaps with the old ontological state, because Alice's and Bob's settings a and b are classical.

are enough to modify the ontological state the entangled photons were in when emitted by the source.

More concretely perhaps, Alice's and Bob's settings can and will be correlated with the state of the particles emitted by the source, not because of retrocausality or conspiracy, but because these three variables do have variables in their past light cones in common. The change needed to realise a universe with the new settings, must also imply changes in the overlapping regions of these three past light cones. This is because the universe is ontological at all times.

References

K. Zuse, *Rechnender Raum,* ed. by A. German, H. Zenil (Braunschweig: Friedrich Vieweg & Sohn, MIT Technical Translation AZT-70-164-GEMIT, Massachusetts Institute of Technology (Project MAC), Cambridge, Mass. 02139, 1969) (Trans. *Calculating Space*)

E. Fredkin, An introduction to digital philosophy. Int. J. Theor. Phys. **42**(2) (2003)

C.A. Feinstein, Why do we live in a quantum world? Phys. Essays **30**(57) (2017). http://arxiv.org/abs/1607.0388

G. 't Hooft, *The Cellular Automaton Interpretation of Quantum Mechanics*, in Fundamental Theories of Physics, vol. 185 (Springer International Publishing, 2016). eBook ISBN 978-3-319-41285-6, https://doi.org/10.1007/978-3-319-41285-6, Hardcover ISBN 978-3-319-41284-9, Series ISSN 0168-1222, Edition Number **1**, http://arxiv.org/abs/1405.1548

J.S. Bell, On the Einstein–Podolsky–Rosen Paradox. Physics **1**, 195 (1964); J.S. Bell, *On the Impossible Pilot Wave*, in Foundations of Physics, vol. 12, p. 989 (1982); J.S. Bell, *Speakable and Unspeakable in Quantum Mechanics* (Cambridge University Press, Cambridge, 1987)

J.F. Clauser, M.A. Horne, A. Shimony, R.A. Holt, Proposed experiment to test local hidden-variable theories. Phys. Rev. Lett. **23**(15), 880–884 (1969). https://doi.org/10.1103/Phys.Rev.Lett.23.880

L. Vervoort, Bell's theorem: two neglected solutions. Found. Phys. (2013). https://doi.org/10.10701-013-9715-7, arXiv:1203.6587v2

Phenomenology, Freedom, Causality, and the Origin of Western Civilization

Abstract Every form of culture—scientific, philosophical, humanistic, religious, etc.—believes that "things" become, transform, that is, "things" occur one after the other according to the scheme of "first" and "then". Besides, every form of culture believes that the succession of "things" appears, i.e., belongs to the manifestation of the world. But since the Greek philosophers, this succession has been understood in two opposite ways: (1) as a process where, instead of the things that actually happen, other, different things could have happened; (2) as a process where this possibility does not exist and it is necessary that things happen in the way they happen, so that the "first" totally determines the "then". In relation to human decisions, the first of these two ways is the foundation of "free will"; the second way, called "determinism", denies the existence of "free will". Modern science has inherited this contrast. Our main thesis here is that neither one nor the other of these two ways of conceiving the succession of things in the world can be something observable, testable,

© Springer Nature Switzerland AG 2019
F. Scardigli et al., *Determinism and Free Will*,
https://doi.org/10.1007/978-3-030-05505-9_3
49

or ascertainable. Determinism and free will are theses that therefore should (although I show in my writings that they cannot) be founded on conceptual structures different from experience. A second aim is to show the following two things: (1) The opposition between these two perspectives exists within their essential solidarity, i.e., the fact that they are two aspects of the same soul: the belief that things come out of their non-being and return to it. This is the dominant soul of Western civilization, and now world civilization, and therefore also of science. (2) These two conflicting perspectives cannot be investigated experimentally, and they cannot appear in experience; and their shared soul has the same characteristic. Experience does not show that things come out of non-being and return to it; the central thesis of my philosophical inquiry is that every state of the world, every part of every state, the content of every instant, every entity, every event, is eternal: it is impossible that they should be otherwise. Western civilization appears therefore to be the history of nihilism. As a third point, we intend to show that, within the history of nihilism, the philosophy of the last two centuries has provided a foundation for the idea that determinism is destined to succumb and free will is destined to prevail, thus making possible the scientific and technological domination of the world. This will lead to the theme of non-nihilism.

I

Phenomenology is a concept developed in philosophical contexts, but it is also an attitude that lies at the heart of scientific knowledge itself. "Phenomenology" is a term deriving from the Greek word *phainomenon*, "phenomenon", which in turn comes from the verb *phainesthai*, meaning the "manifesting" of things, their "appearing", "showing", "being in light" (*Phainesthai* is in fact a derivation from *phos*, "light"). And at the root of scientific method is experiment, experimental observation of what lets itself be seen—the *phenomenon*, precisely. Galileo states that science is made up of "sensible experiences and necessary demonstrations". Experience is the field of the observable. Experiences have to be "sensible", because they can also be meaningless (with respect to certain criteria considered reliable). The experience of primitive man, for example, is not a form of relationship with the world on which science would rely for the verification of its theories. For science, the archaic experience (although not only) is too full of impulses, fantastic images, and dreamlike elements, which alter the disclosure of reality, i.e., the "facts".

However, the concepts of "phenomenon" and "experience" have been radically questioned. This has happened in both the philosophical-epistemological and the scientific sphere. Nietzsche's principle that "there are no facts, but only interpretations" has been transferred in the theorem that "the facts are loaded with theories" (K. Popper); and above all in non-deterministic physics, it has been shown that observation can substantially change the observed reality. The meaningfulness of *pure* experience is a mirage. Yet these criticisms of the concepts of "phenomenon" and "experience" cannot prevent the disclosure, the appearing, the manifestation of things from being reconstructed at a higher level. For if the facts hide or even disappear behind the interpretations, and reality behind the modification of itself, operated by the observer's presence, interpretations and modifications will nevertheless appear and manifest themselves. The "thing" loses the semantic simplicity in which it was naively concealed (and for which tables, windows, mountains, factories, cities, and financial flows show the properties that are commonly attributed to them), and it appears as an interpretation or modification of a reality that remains hidden (so that tables, mountains, etc., appear as contents of the interpretation or as modifications of the reality, i.e., modified reality). So, what appear or manifest themselves are precisely those (complex) things usually called interpretations and modifications.

The process of their occurring also appears—and this is the world. Science (and more generally, human life) is based on the manifestation of the world. It could not take a single step without it. But science rarely turns back to investigate the meaning of what it relies on (and the same generally happens in the various pursuits of human life). The world's manifestation includes everything that appears. If something does not belong to the manifestation of the world—if it does not appear in this primary sense of appearing—it cannot appear in any kind of experience or observation. This can also be said when science reveals the illusory character of experience: if the Sun's motion did not appear, science could not claim that it was an illusion.

However they are understood, "things" happen according to the scheme of "first" and "then"; and their occurrence *appears*, namely belongs to the manifestation of the world. But since the philosophical thinking of the Greeks, succession (motion, becoming, change) has been understood in two contrasting ways: on the one hand, as a process where, instead of the events that happened, other, different ones might have happened (and this possibility may concern either the totality or a part of what happens); and on the other hand, as a process where this possibility does not exist at any stage of the process and it is necessary that events happen as they happen, so that the "first" completely determines the "then". In relation to those events that are human decisions, the first of these two ways is the foundation of what we call "free will", while the second, which we call "determinism", denies the existence of "free will" (one of the primary meanings of freedom). Modern science has inherited this contrast—and it could be said, for example, that Einstein supported a deterministic perspective, whereas Heisenberg opened the door to the assertion of free will.

The *thesis* we intend to show in what follows is that *neither* of these two ways of conceiving the succession of things in the world *can be a content that appears*, i.e., that belongs to the manifestation of the world; and therefore *neither one nor the other* is something observable, testable, or ascertainable. If this statement is true, it implies that, from the point of view of science, both the affirmation of free will and the affirmation of determinism are theoretical constructions, assumptions that are expected to be confirmed or falsified by experience. But the confirmation of determinism would be the occurrence of a case (or several cases) where determinism appears, and the falsification of it would be the occurrence of cases in which free will appears; and likewise the confirmation of free will would be the occurrence of cases in which free will appears, and the falsification of it would be the appearance of cases where determinism appears. But if the thesis we intend to affirm (i.e., the impossibility that both the determinism of events and the free will appear) is true, then none of those confirmations and falsifications will be possible.

The consequence of all this is that neither the point of view of determinism nor the point of view of free will are of a scientific nature. Whatever cannot have any confirmation in experience—we repeat—cannot be of a scientific nature. And we intend to show precisely the absence of *any* confirmation here. By saying that *neither of the two* opposing ways of conceiving the succession of things can be a content that appears, we intend to say in fact that it is impossible to experience even a single case where what happens could not have happened, or where what is happening is totally determined by the past.

Clearly, there remains the problem of the consistency and, we may say, of the "truth" of the thesis that we intend to support here. This is also a decisive problem—or it belongs on the general level of what is decisive. But the discussion of that problem would lead us too far from the theme that concerns us here. However, we shall come back briefly to this issue in our concluding remarks.

II

Free will *is not* a particular case of the *simple* "contingency" of events, that is, of their occurring, not permanently taking up the world scene, and preventing other events from happening. The simple "contingency" is the succession that leads from "before" to "then"; and, as we have said, the occurrence of events and things *appears* (and is therefore testable). Hence, the particular case of "contingency" where, for example, you are initially seated and then you decide to stand is something that appears. However the simple "contingency" is not the (much more complex) *situation* where, *instead* of the events that happen, other, different events could have happened. In fact, this is the situation where, *instead* of the events that *appear*, others might have appeared, but did not in fact.

But the appearing of things in the world cannot say anything about what could have appeared; i.e., it is impossible for the appearing of things to include what does not appear (and that it does not appear exactly because it is what might have appeared). This situation is the *complex contingency* of what actually is, or appears, but could not be, and could not appear. Free will is the complex contingency of human decisions. It turns out from what we said that the non-appearing of free will is not a "fact" that could be replaced by the opposite fact, but is a *necessity*. It is, therefore, *impossible* that free will appear as a simple "fact". If we affirm that free will appears or may appear, we affirm that the dimension of what appears includes what does not appear, namely that something is not what it is: that X is non-X. (On the full meaning of necessity and impossibility, I must refer here to my other publications.)

Additionally, if one argues that free will is about the existence of things but not about their appearing (in the sense that things in themselves could possibly not exist, and those that do not exist, could exist—whence for them there is not the impossibility that the appearing of things in the world include what does not appear), this thesis acknowledges what we intend to affirm here, namely that complex contingency and free will are not a content that appears, that is, they are not an observable, testable content. Kant's position is congruent with this thesis. For Kant, freedom belongs to the noumenal world of things in itself; while experience has a deterministic character, and thus does not attest to the existence of free will or, in general, what we have called "complex contingency."

On the opposite side, this contingency and free will is something absolutely obvious. For Descartes, nothing was more certain (*nihil certius*) than free will. The Jesuit Rodriguez Arriaga was so convinced of the obviousness of this that he referred to those who deny free will by declaring Aristotelically that they should not be dealt with by reasoning, but by beating (*non ratione sed fustibus agendum est*).

After explaining the reason why what we call "complex contingency", and therefore free will, do not appear and are not observable, we must now explain why the determinism of events cannot be something that appears and is observable or testable.

The deterministic perspective argues that events are *totally determined* by the previous state of the world. Such *determination* actually produces them; that is, the former state of the world is their *cause*, which *necessarily* produces them. Determination is *total* in the sense that it is also sufficient to produce the events. The deterministic perspective is the most radical statement of the "principle of causality". This principle is dominant in the sciences. Quantum physics is the exception. But just because it is necessary, not even the causal connection can be something that appears, something observable or testable. When, according to the principle of causality, a state of the world produces the next state, i.e., in the act where the cause actually produces the effect, the cause does not act as something which may not be followed by the effect, i.e., the effect is not something that happens but that could have not happened; rather, the effect is necessarily produced, it is necessary that it begins to exist.

But philosophical considerations have long since established that experience cannot attest to any necessary relationship between the events that appear in it, hence not even the causality relation. In fact, a relationship between X and Y is necessary only if it exists not only in the content that appears, but also in those that will appear and have appeared, and also in those that do not appear and that are not experienced, and even in those that will never appear. Which means that it is impossible for experience to show what does not belong to the experience. And if one affirms the existence of a necessary relationship that exists only within a particular dimension of reality (for example, within the present experience), this relationship is open to the possibility of being denied by the coming into being of new states of the world, or by a different configuration of experience: it is not a necessary relationship.

From what has been said, it follows that determinism and free will are theses that must be based on conceptual structures other than experience. And attempts in this direction have been made throughout the history of philosophical and scientific thought.

But *can* those arguments be well founded? The answer is *negative*—it is necessary that this be so, although in the present paper it is inevitable that the *roots* of that answer be kept in the background: elucidating them here would take us too far from the present issue. For their clarification I refer to the whole of my philosophical discourse. But we may make a few remarks about the relationship between what has been said so far and the *origins of Western civilization*, thus taking a few steps in the direction of those roots.

III

In the history of peoples, philosophy is the advent of the will to evoke knowledge that cannot in any way be denied. It is therefore a negation of myth, faith, and opinion. But a piece of knowledge can be absolutely undeniable only if its content is not a part beyond which other parts exist. If, in fact, other parts exist, they may be governed by different laws, or obey no law, and the claim that that knowledge is absolutely incontrovertible would be futile. To be absolutely undeniable, it is therefore necessary that its content is not a part, but the whole. However, the whole must necessarily be the totality of *what it is*: all *entities*, and, to begin with, the entities that *appear*. For the first time philosophy is thinking about the extreme opposition of *being* and *nothingness*: not only because the totality is that beyond which there is nothing; but also because a being that appears does so as something that comes out of its own not being (i.e., from its being nothing), and returns to its own not being, allowing other entities to begin to be.

This occurrence of the entities which appear was then understood either as a deterministic process, or as what we have called "complex contingency", one aspect of which is free will. But the opposition between these two perspectives lies within their essential *solidarity*: they are two aspects *of the same soul*, namely, the conviction that beings (those that appear, or all beings) come out of their non-existence and go back to it.

The philosophical tradition believes that the becoming of entities can exist only if there *exists* an immutable Being; while in its essence the philosophy of the last two centuries shows that becoming can only exist if *there is no* immutable Being. However, this opposition, which is a contrast between forms of life and society, is also subordinated to that essential solidarity; it also takes place within the same soul, that is, the soul of the entire Western civilization. And since the fundamental categories of the West are dominant almost everywhere, it is now the soul of human existence on Earth.

Science also develops within this soul. Remaining on the contrast between determinism and contingency-free will, for the deterministic standpoint, the cause predetermines the effect, and yet the effect is not entirely contained in the cause (otherwise the future would already be fully realized in the cause). Hence the future effect, despite its anticipation in the cause (despite its being already, because anticipated), *is not yet* (it is still nothing); and the effects produced in the past *are no longer* (now they are nothing), even if, as they continue to determine everything that comes to exist after them, they are still. If we eliminate the *not yet being*, and the *no longer being*, we eliminate the whole of scientific knowledge. (For example, the amount of energy in the Universe is constant, but the different shapes that the Universe assumes during its expansion—that are not "nothing"—*were not yet*, i.e., they were nothing, before they appeared, and *are no longer*, they return to being nothing, when other shapes replace them. Or again, as another example, before the heat of a hot body moves to a colder one, in the cooler body that heat *is not yet*, and when the heat has passed into it, *it is no longer* in the body that was warmer. Or again, before light reaches bodies, their brightness *is not yet*, and *it is no longer* when they are no longer illuminated.)

In the simplest but most fundamental way in which it can be presented, Heisenberg's uncertainty principle states that any state of the world does not predetermine the next state in any way. The prediction of the future can therefore only be statistical or probabilistic. We may even say that the future is unpredictable. So what happens could have not happened. In this way, even decisions taken could have been not taken. Just because of this, a free decision is unpredictable. The non-deterministic world makes possible free will.

But the fact that a state of the world does not predetermine in any way the next one means that, when the first state exists, the next *is not yet*, it is still nothing; and when the next state exists, the previous one *may be no longer*, it may be canceled (because otherwise there would be a necessary link between the two states, which is precisely what is excluded by the principle of uncertainty).

IV

The belief that entities come from their not being and then return to it is the ground in which all the conflicts in the history of Western civilization have grown—their common soul, as we put it previously, and therefore also the soul that is common to determinism and free will. All forms of production, creation, and destruction are possible only on the basis of that belief. But now it should be noted (referring briefly to what I have explained in depth elsewhere) that it is not merely these two conflicting attitudes that *do not appear* and *are not able to appear*, but *also their common soul*, the same ground in which the whole story of the West has grown.

In a more or less explicit sense, the various forms of Western wisdom regard that soul instead as the original absolutely undeniable *evidence*, that is, as what is maximally observable and testable, as the content that undoubtedly appears even when things are no longer conceived according to the parameters of the common sense, but as interpretations (see Sect. I). Even when the illusion of becoming is affirmed, conceived in the indicated sense, it is affirmed precisely because we are convinced that becoming appears.

Those conflicting attitudes are not observations, as has been shown, but theories; and it is a *theory* and not a content that appears, even the belief that the occurrence of entities is their oscillation between being and not being. This theme is fundamental, but here we shall confine ourselves to just a few notions.

When an entity is not yet (when it is nothing), it is impossible for it to appear (be observable or testable): that appearing would be the appearing of nothing, that is, a not appearing. (To be precise, if it is possible to predict anything about an entity that is still nothing, it is impossible that in the prediction it will *look anything like* it will appear when it does come into existence.) And when a being is no longer (it is nothing once again), it is impossible for it to continue to appear (i.e., it is impossible for it to keep appearing *as* it appeared before). But if, when it is not yet, an entity cannot even appear as it will appear when it does come into existence, then it cannot be the appearing of the world, observation, and experience that certifies that it is not yet and it is still nothing. Likewise, when an entity is no longer (when, for example, a city is destroyed) and therefore cannot even appear as it appeared before it was no longer, then, again, it cannot be the appearing of the world that shows that it is no longer and is now nothing.

That entities come out of nothing and return to it cannot be affirmed by something like experience or observation, and yet this is what is indeed affirmed by the *fundamental theory* of Western civilization. (On the other hand, we can show that the Orient is the prehistory, the incubation, of the West.)

What is the foundation of this theory? Or will we have to admit that this theory cannot have any foundation because it is indeed *an extreme alienation of truth*? And that affirming that entities oscillate between being and nothingness means identifying *what it is* and nothingness? That is, that it means *wanting* the annihilation of being?

V

These questions underline the central thesis of my philosophical inquiry: that every state in the world, every part of every state, the content of every instant, every entity, and therefore every event, is *eternal*. To be eternal means that it is *impossible* for it not to be.

This impossibility is not ascribable to any other form that impossibility has taken in the wisdoms of the West, therefore not even to the meaning that scientific knowledge confers upon this word. It is, however, interesting to note that there is a way of interpreting the theory of relativity as an affirmation of the eternity of every state in the world, and that Einstein did not oppose this interpretation.

In his *Intellectual Autobiography* (1976), Karl Popper writes, referring to Einstein: "The main topic of our conversations was indeterminism. I tried to persuade him to abandon his determinism, which in practice consisted in the idea that the world was a closed universe, of Parmenidean type, four-dimensional, in which change was a human illusion, or something very similar. (He agreed that this was his opinion, and discussing it, I called him 'Parmenides')." Einstein accepted the interpretation that embodies the theory of relativity in the philosophy of Parmenides. Later he would show less confidence in his adherence to determinism; but it does not seem that his Parmenidism had faltered.

Popper's intention of converting Einstein to Heisenberg's indeterminism was already an attempt to reconcile what appears to be irreconcilable: the theory of relativity and quantum mechanics. An attempt that in physics has attracted much interest and is still ongoing. Of course, the discussion between Popper and Einstein is not about the 'historical' Parmenides, the one discussed by Plato, Aristotle, and Hegel. The 'historical' Parmenides supported the eternity of pure *Being*, not of *beings*, or entities, i.e., of things that are. Almost from the beginning, my writings have instead shown the *necessity* that each thing-that-is is eternal, where "thing" is to be understood in the widest possible sense, including every event, relationship, and gradation, and thus every space-time event, i.e., also (but not only) the events of the four-dimensional chronotope of the Einsteinian universe.

And this *necessity* is the *impossibility*, mentioned above, for any being not to be eternal. It should be noted that the 'logic' under which the theory of relativity states that everything is eternal is essentially different from the necessity that my writings address. Also because science nowadays—including the theory of relativity—recognizes the hypothetical and provisional character of its own theses, even the most "confirmed" ones. And it is in this sense that philosophy in its essence can go beyond science (which, on the other hand, is now concerned about power, not about "truth"). The fact remains, however, that the Parmenidean thesis of the theory of relativity *sounds* identical to the thesis of my writings, that every being is eternal—even if the foundations of the two theses are radically different, so that the theses themselves *are* different.

And that's not all. Popper and Einstein agree that for Einstein, as for what they consider to be the historical Parmenides, the "experience" (appearing) of change is illusory. In my writings on the other hand it is shown—as mentioned above—that although the "experience" shows the *variation* of its contents, it does not show that specific sense of change which for Einstein and Parmenides is illusory, and which is the "becoming" as the coming out of entities from their not being and their returning back into not being.

For Einstein, the experience of this transition from not being yet to being no longer is illusory. Since in my writings it is shown that the "experience" does not attest and cannot attest to the transition from what is not yet to what is no longer, these writings thus show that the "experience" is *not* illusory.

Therefore, since every being is eternal, the *variation* of experience is the *appearing and disappearing of eternities*. Of course, the sense and the structure of the appearing is extremely complex (here we content ourselves with merely evoking it), but it would have allowed Einstein to respond to Popper's criticisms.

In his discussions with Einstein, Popper often proposes the metaphor of frames and their projection. Every being is like a frame of a film; the frames are all coexistent and immutable and it is their projection that gives the illusion of motion. Projection corresponds to the illusory experience of motion. Einstein accepts the metaphor. But Popper objects (*Postscript to the Logic of Scientific Discovery*, Vol. II, Sect. 26): "If we experience successive images of an immutable world, then one thing at least would be genuinely changeable in this world: our conscious experience. A cinematographic film, although presently existent [in its entirety], and predetermined, has to *pass* through the projector (that is, relative to ourselves) to produce the experience or the illusion of temporal change. [...] And since we are part of the world, there would be a change in the world—which contradicts Parmenides' view." And also, I observe, the opinion of the non-historical Parmenides, who affirms the eternity of all the many beings in the world—where, on the contrary, the historical Parmenides of Plato, Aristotle, and Hegel denies the existence of a multiplicity of beings and, in addition to becoming, also qualifies such multiplicity as illusory.

However, regarding this criticism (and others), Popper reports: "Einstein said he was impressed and did not know how to answer" (*ibid.*). This attitude of Einstein's is noteworthy, because it shows that the theory of relativity involves not only the physical world, but also our "conscious experience". (On the other hand, his *realism* contradicts his eternalism, because if reality exists even when there is no "our experience that is conscious of it", then this experience is something—an entity—that can be nothing, hence is not eternal.)

Einstein does not know how to respond to Popper's criticism because—like Popper and like every wisdom of Western civilization, and now world civilization—he in turn feels that change is to be understood as a coming out of things from their not being, and a return to their not being (howsoever that change is conceived, i.e., as real or illusory). So even Einstein is forced to conceive in this way of the *passing*, or *flowing*, of the film through the projector. Even for him, in our "conscious experience", which he regards as illusory, things go from not being to being and vice versa. And nor does he realize that this passing does not appear and cannot appear.

But outside the untruth in which that sense of passage consists, what we are saying is that the change is the appearing and disappearing of the eternal beings. And since also their appearing is an eternal being, *their appearing also* appears and disappears.

VI

My writings have already taken into consideration an objection that would later be addressed to them. This objection goes as follows: if and since change is the appearing and disappearing of entities, will you not have to say that *at least* their appearing comes out of nothing and returns to it? The *appearing* corresponds to the projection of Popper's film.

But this time we may answer by saying that, *just as* the eternal in which this voice, or this shadow, or this memory consists begins to appear and now no longer appears, *so* the eternal in which the *appearing* of this voice, of this shadow, of this memory consists begins to appear and now does not appear any more. The manifestation of the world—namely the appearing of the world, which is the present totality of what appears—is the *place* to which all that begins to appear comes, and from which all that no longer appears exits. And this place is not only eternal itself (like every entity), but it is also impossible for it to start to appear and no longer appear, since this beginning and this ending is the beginning to belong to that place, and the not belonging to it any more.

Since the leaving out of nothing and the returning to it does not appear (see Sect. III), the affirmation of the eternity of every entity does not deny, and is therefore not denied, by the manifestation of the world. In contrast to Einstein (when he discussed with Popper), who could not answer the objection that Popper put to him: not even Einstein (and not even any other wisdom of the West) could find the words to indicate the genuine sense of appearing. What appears in the place where the appearing of the world is, is not the coming out from nothing and the return to nothing of entities, but rather the appearing and disappearing of those entities; and even the entities that appear and disappear are eternal. They are eternal because—and even here we are confined to a simple remark—if they become nothing or come out of nothing, one should affirm a time when *they*, which are *entities*, are *nothing*; a time when the entity and the non-entity (the nothing) are identical. The genuine essence of *nihilism* consists in a faith in the existence of this time. This is a simple remark indeed, considering the decisive conceptual weight of the theme. (See, however, E.S., *The essence of nihilism*, Verso Books, London—New York, 2016.)

VII

Since both the deterministic theory and the free will theory, in their various forms, hold that entities oscillate between being and not being—that is, they *share* their soul (see Sect. III)—the opposition of the former theory to the latter (and vice versa) can only be *weakened* by this commonality.

This is not the case when, going beyond the beliefs of Western civilization, we can find the words to describe the true sense of nihilism, and hence the eternity of entities as entities (the true sense of which differs essentially from every form of eternity that, in the wisdoms of Western civilization, is built upon the notion of becoming). Indeed, if and since every entity is eternal, no entity can be without every other entity being—and thus without the existence of that entity which is the appearing of any other entity. This means not only that it is impossible that what is might not be, but that it is also impossible that what appears might not have appeared. Free will *cannot* take refuge in this second possibility, once it has been established that the first possibility cannot concern free will. If, in fact, what appears might not have appeared, the totality of entities could have been without that dimension of the entity which is what appears. But it has come to light that this is impossible. So it is necessary that everything that appears, appears, and appears as it appears.

In the history of the West, Aristotle spells out once and for all the sense of "responsibility": if the one who acts is the cause of his action, and is aware of the purpose he intends to reach, the society in which he lives takes account of this circumstance and reacts in accordance with the good or the evil which it considers to have received from such action. For a large part of Western culture, this 'being the cause of one's own action' is equivalent to enacting one's volition as free will. "Responsibility" is the content of a faith (conviction) that belongs to nihilism, namely to the faith (conviction) that things come out of their not being and return to it. This conviction implies in fact that there may be something like "will"; this is the conviction which, believing that it is the cause of its own action, believes also that it has the ability to bring things out of their non-existence and return them to non-existence. And it is believed that "morality" and "ethics" can only exist when the "will" is so conceived.

However, this way of conceiving of "responsibility" *minimizes* it. Precisely because the decision is free, it is thus unconstrained by the whole history of the person that acts. *Except for free decisions*—which are a minority if we consider the way people generally lead their lives—people are not "responsible" for what they do during their lives, in the view of those who support the idea of "responsibility" and free will.

But the story of the will belongs essentially to the history of nihilism. It is the story of Extreme Folly, of the Dream farthest from the authentic Vigil. On the other hand, the Dream of Extreme Folly and all its contents are not nothing—and neither are the contents of the dreams of everyday life—whence the story of the will is also a manifestation and disappearance of eternities. Thus, all the contents that form the story of a man are also eternal. Each of them is therefore in a necessary relationship with every other entity, and thus also with every other moment of that story. This means that the *whole story* of a man is, in a specific way, that without which his decisions could not exist. (In a specific way, since, in a generic way, the totality of the entities is that without which his decisions may not exist.)

Outside the nihilism of the West, then, it can be said that the true sense of the eternity of being as an entity and its implications, far from eliminating or minimizing "responsibility", makes it extreme, inside the Dream constituted by our acts and decisions.

VIII

Acting and deciding are first and foremost a way of designing the future by prefiguring it, according to the sense of the world that is present in them. The meaning of the world brought to light by philosophical thought lies at the heart of the increasingly widespread set of practical-decision/cognitive forms which has transferred from Greece to Europe and now dominates the Earth—the set we refer to as "Western". However, despite their common soul—nihilism—the deep rift that exists between this philosophical tradition and the last two centuries of European culture has now come to the fore.

From the Greeks to Hegel, this tradition affirms the existence of absolute truth, which culminates in the affirmation of the existence of a divine and immutable being, upon which the becoming of the world is founded. And the immutable being is reflected in the immutable configurations of the world that regulate human life: morality, ethics, and natural law understood as conforming to the truth by the individual, the state, and the legal system. The dominant attitude of monotheistic religions is concerned about not being in conflict with philosophical truth. Later, following this tradition, absolute and totalitarian states have tried to have, or at least appear to have, the same attitude. Without truth, beauty is impossible. Modern science has taken its distance from Aristotle, but intends to be the true philosophy of nature, which recognizes, above itself, the "first philosophy" as the undeniable knowledge of all entities, and of the immutable being. Market laws are also considered to be "natural laws", that is, undeniably true; and Marxist communists also intend to possess that absolute truth which in their view cannot be recognized in capitalism.

But over the last two centuries, this great picture of the Western tradition has been slipping down below the horizon. The undoubted visibility displayed today by religions is the swan song of the Western tradition. No one can turn their back on the current that is dragging the world away from that tradition. The most visible aspects of this current are events like the French Revolution, the Soviet Revolution, the world wars of the last century (i.e., the victory of democracies over the absolutism of the central empires and true socialism), and the transformation of the customs of advanced 20th century societies.

But these are also events that can be judged as the simple success of one force over another, considering that it would have more *rights* for world dominion, because its dominion would be the dominion of truth. Yet this judgment is arbitrary, because below the surface of what appears to be the winning current flows the criticism that the culture of the last two centuries addresses to tradition. And above all, the symptoms of an unstoppable flow of change. For example, the transformation of democracy from the Greek form, where freedom was inseparable from truth, to a procedural democracy, which does not uphold the truth of its laws but considers the will of the majority of voters as the sole criterion for their value. And again, the resulting replacement of natural law by positive law, where the winning historical forces dictate the content of the laws; and the liberation of art from its subjection to truth and beauty, which must together be true, a liberation in which art becomes "abstract art", that is, abstraction, and hence the negation of the beautiful so understood.

And just as we speak of "abstract art", so we can talk about "abstract science" (hence also abstract politics, morality, etc.). This relates on the ability of scientific knowledge to question things, that is, "to abstract" from the absolutist conception of science in which Galileo could assert that, although man knows only some of the mathematical truths, yet he knows them in their absolute truth, namely, in the way in which they are known to God (who knows them all). We may list some examples of how modern science overcame the absolutizing attitude that it so long maintained after its birth: the emergence of non-Euclidean geometries (which no longer allow Euclidean geometry to be regarded as absolute truth); the discovery by Gödel of the impossibility of being sure of the absence of contradiction in mathematical knowledge; the concept of statistical and probabilistic prediction introduced by quantum mechanics; and in general, the emergence of a belief in the hypothetical and therefore falsifiable nature of science.

IX

But at this point, the abandonment of the absolutist conception of truth either recognizes that it cannot have the character of absolute truth, and therefore has only a hypothetical and falsifiable value, or presents itself as an nth-form of that naive skepticism that Greek philosophy had already rid itself of. How then is it possible to abandon the faith in the existence of absolute truth, which has the character of absolute truth, but is not in itself a faith? Much of the philosophical knowledge of our time is still unable to answer this question.

However, we know that throughout the history of Western culture the existence of coming from not being and returning to it is considered as an absolutely "obvious" and indisputable truth: *both according to* the tradition and *according to the negation* of tradition. The philosophical wisdom of the West—we are about to show—is capable of responding to this question by appealing to this "obvious truth" (even acknowledged by those who, like Parmenides or Einstein, affirm that the experience of becoming is an illusion, since the content of this experience is also surely understood by them as the oscillation of entities between being and not being).

In the last two centuries of philosophical knowledge, becoming, understood in this way, has become the undeniable foundation for the denial of every other undeniable truth, and above all, the undeniable truth that thinks it can show the existence of the divine immutable being that sustains the existence of all entities. And all this without the abandonment of tradition presenting itself as naive skepticism. That is to say, without such abandonment, *which is the most coherent form of nihilism*—and which, on the other hand, does not see and cannot see itself as nihilism—presenting itself as naive skepticism.

The abandonment of tradition described here lies in the *philosophical subsoil* of our time. On the surface, the philosophy of the last two centuries expresses rather the *need* to bring that tradition to an end. It tends to remain a faith. Few inhabit the subsoil, and in any case one has to be able to get down there. Even with regard to this central theme of Western history, one has to confine oneself to certain notions, especially if one intends to point out the conceptual configuration that is *shared* by the inhabitants of that subsoil.

The essential subsoil of nihilism (that is, the most rigorous form of nihilism) denies any absolute truth that is *different* from the absolute truth of the becoming of entities, and which is precisely the absolute truth of tradition. The essential subsoil of nihilism keeps in mind that this absolute truth would like to be the law of *every* entity; not only of the presently existing entities, but also of those which are no longer, and of those which are not yet. What is no longer and what is not yet—says the subsoil—does not belong to a kingdom that is capable of escaping from the absolute truth, because it is the prerogative of such absolute truth to establish the sense of being now, and being no longer, and not yet being; and also because if that kingdom existed, it could break forth and overwhelm the realm of absolute truth, which therefore would no longer be such.

And what is said of the absolute truth must also be said of the existence of the immutable, and therefore of the eternal God, which the absolute truth considers it is able to prove. Absolute truth is the absolute law of knowledge that shows the absolute law of the immutable being; since the immutable, God, is in turn the inviolable law not only of what is, but also of what is not yet and what is no longer. The God of the Western tradition does not permit a region threatened by a past and by a future that can escape the divine law. God is the lord of every entity and of all time. He is indeed the cause of time because He is the cause of becoming. So it is on the basis of the evidence of becoming—that is to say, coming out of nothing and returning to nothing—that absolute truth, as "metaphysics", believes it can demonstrate the existence of the immutable, which is the foundation of the existence of every entity, and of the same absolute truth.

But, just because of this very feature, the Legislation of Everything, constituted by the unity of the absolute truth and of God, transforms into an *entity* the *nothing* of what is not yet and of what is no longer. Under such Legislation, this nothing becomes a listener and a subject, something that is held inside the existence. In fact, concerning what is not yet—and therefore is nothing—such Legislation requires it to be something which can exist in the future, but which in the meanwhile is that kind of being that is the "possibility" (Aristotle would have said a "potential"), and which, when it comes into being, will have to comply with certain rules (such as the principle of non-contradiction, or the dialectical method, or being in relation with other entities). Concerning what is not "yet", and is therefore nothing, that Legislation requires it to be that *positive* that is the "*not yet* being"; and concerning what is no longer, it requires it to be that *positive* that is the "*no longer* being". This Legislation therefore imposes an adaptation to the rules of truth that have allowed it to be, and which also guide what is not yet, and which allow the memory of what is now nothing.

As a subject and a listener of the Law of absolute truth, therefore, what is nothing can no longer be that nothing that the metaphysical tradition considers to be impossible to eliminate, namely, without which the becoming from nothing and the becoming nothing would be impossible. If there is such an immutable and eternal Legislation, then becoming is impossible; but becoming is the supremely undeniable and obvious fact (both for the tradition and for the subsoil that destroys that tradition). Therefore, the existence of every absolute truth and of every immutable being or God is impossible. And because of that, any necessary relationship is impossible, *necessity* being the essential character of absolute truth.

This, in short, is the conceptual configuration that is *common* to the inhabitants of the essential subsoil of our time, where nihilism reaches its most rigorous form. To name a few such thinkers: Nietzsche (who did not want to exhibit the *true* power of his thinking), Giovanni Gentile (who was not studied on the pretext of his adherence to fascism), and Giacomo Leopardi (whose philosophical thinking anticipated Nietzsche's, and who has only recently begun to receive the recognition he is due). And the list struggles to go beyond these names.

X

But the consequences of what that subsoil can discern are relevant for the relationship between determinism and free will. These two opposing conceptual dimensions share the soul of nihilism, but within it and its history, the faith in determinism is destined to succumb, while the faith in free will would prevail. Determinism, in fact, is supposed to provide a necessary link, according to the meaning of *necessity* that belongs to the absolute truth of the Western tradition. The necessity for events to occur according to the rules of determinism is in fact valid not only for the entities that are present, but also for the entities that are not yet present, and for those that are no longer. So, even here, their being nothing becomes a listener and a subject of the deterministic order, and therefore what is nothing is transformed into an entity; and the vanishing of nothing is the vanishing of that coming out of nothing and returning to nothing which, even determinism, like the whole of Western culture, considers to be the original evidence and absolutely undeniable.

Faith in free will (and in general what we have called "complex contingency") does not prevail because it is testable and observable (see Sect. II), but because it is implied by the impossibility of determinism and of any absolute truth of the tradition. In this sense, quantum physics and the uncertainty principle are destined to prevail over the vestiges of determinism still present in contemporary physics. It is no accident that Heisenberg traces the concept of "probability wave" back to the Aristotelian concept of *dynamis*, that is, "opposites in power" (and therefore even opposing decisions in power), each of which could be realized instead of the other.

The consequences of what the essential and essentially philosophical subsoil of our time can discern are ultimately decisive in relation to the ability of techno-science to dominate the world.

The voice of the subsoil is still heavily overwhelmed by the din at the surface, where, on one side, "God is dead" tends to become, as Nietzsche foresaw, a song for organetto or, at best, a simple need to abandon the essential (philosophical) tradition of the West, considered as something we should no longer linger over. But on the other side, this tradition can warn those who want to eliminate it, and above all scientific knowledge, by saying that scientific knowledge is, and recognizes itself to be, a body of hypothetical, provisional knowledge, and that there are in fact inviolable limits—shown by the absolute truth of philosophical-theological-metaphysical knowledge—beyond which techno-science must not go, because it has no right to do so.

But when and where the action of man, and above all techno-science, can hear the voice of the essential subsoil, and feels that the existence of those inviolable limits is in fact impossible, then that voice opens the way for technology and, by telling technology that its desire for power does not have limits, the voice *really does give* technology the power that it cannot have as long as it thinks it serves only as a means to achieve the purposes that tradition presents to it. Just as a man does not run as long as he does not feel he has legs. The world is moving towards an era of technological dominance, made possible because the subsoil of current philosophical thinking now prevails over traditional Western (and even more so, Oriental) thinking. The world is moving, by every means available to it, towards the most rigorous form of nihilism.

Beyond nihilism, *the fate of truth* sees that every form of power, and hence also the power of techno-science, can only be a dream, for if and since every being is eternal, no will of power can modify it, produce it, or destroy it. This is where the great themes of awakening open up.

But it is appropriate to close these considerations by reiterating that the *fate of truth*, that is, the issue of non-nihilism, is not just another reappearance of the absolute truth of the philosophical tradition. The *destiny* of truth is a kind of absolute *staying*. The "*epi-stéme* of truth" is instead the failed *attempt to stay* (*steme) by the content of the philosophical tradition. It *is failing*, we have seen, because it makes impossible the sense of becoming which belongs to the very foundation of the *episteme* of truth. As long as we maintain our faith in not-being-yet and in being-no-longer, it is inevitable that any absolute truth different from the content of that faith should be impossible, and that the only absolute truth, the only undeniable evidence, is precisely that content (which cannot therefore be seen as the content of a faith).

But the fate of truth is the *negation* of nihilism, which conceives as evidence the faith in not-being-yet and in being-no-longer that makes every absolute truth impossible, if different from the content of that faith. The negation of nihilism therefore provides the language *to speak* of the absolute *staying* of the fate of truth. The fate of truth makes possible the language that testifies to it.

Bibliography

I. Kant, *Kritik der praktischen Vernunft*, Werke, Band 7. (Frankfurt a. M., 1977)

W. Pauli, *Theory of Relativity* (Dover Books on Physics, 1981)

K. Popper, *Unended Quest: An Intellectual Autobiography* (Routledge Classics, 2002)

E. Severino, *The Essence of Nihilism* (Verso Books, London, 2016)

B. Spinoza, *Etica* (Edizione critica del testo latino, ETS, 2014)

Grace, Freedom, Relation

The essay seeks to illustrate the contribution of the biblical Christian experience and understanding to the question of the relation between freedom and necessity. With an epistemological approach that is formally theological, it aims to provide a cosmological and ontological horizon of meaning and thereby promote a dialogue between science, philosophy, and theology. According to theological language, grace expresses the meaning and the destiny of reality as being originated, informed, and directed by gift and forgiveness. Freedom is understood not only as the possibility of choice, but as the core expression of what is human, both in being guaranteed and founded upon the grace of God and being fully realized as a freedom allowed by the grace of God; such freedom is realized where it is given in relation to, and as a relation to, another in mutual recognition. It thus becomes possible to draw from the theological concept of creation to propose a renewed

(Translation by Julie Tremblay)

paradigm for understanding the transcendence and immanence of God with respect to the world. The transcendence of God is so transcendent as to express itself in the
most perfect immanence, in as much as reality is created
freely by God, through grace, in the call to freedom.
Therefore, none of the perspectives for the interpretation
of reality are to be considered as absolute, whether they be
theological, philosophical, or scientific. Rather they need
to interact, respecting the specific formalities and level of
each, and listening carefully and without preconceptions
to the reasoning involved in each approach.

A Methodological Premise

In this article I shall seek to express what the biblical
Christian experience and understanding bring to the dialogue between grace, freedom, and relation. The focus is
primarily theological and anthropological, from a Catholic
perspective. In my opinion, it discloses an ontological and
cosmological horizon that promotes and opens the way for
comparison between science, philosophy, and theology.
As a premise, it is useful to recall some methodological
aspects that will allow for a pertinent interpretation and
a fruitful interaction—from my point of view—between
philosophy and theology in the context of this topic.

To simplify, we could say that the reality that surrounds
us and that we live in is seen, interpreted, and to some
extent also made by us from a multiplicity of different
approaches which in no way exclude one another. Rather,
in a certain sense—which must clearly be understood and
managed with prudence and wisdom—they complement
one another.

The first approach (especially today, and in many
ways, if nothing else at least with regard to evidence and

practical usefulness) is the scientific approach, in the modern sense of the term. It involves reading reality according to what many today refer to as "methodological naturalism", which means interpreting reality on the level of its physical, chemical, biological, and psychological expression. Since these involve measurable quantities, they can be checked using the instruments available in each case, *iuxta propria principia*, starting from and according to their own principles, based on the structure and dynamics of each of these levels. This is done using the experimental method, building models based on justifiable hypotheses, making predictions, and objectively checking the results. It is the type of approach used, for example, even if in different ways according to their specific fields of inquiry, by Galileo Galilei in physics and astronomy and Charles Darwin in biology, two names that have revolutionized our way of viewing the world and of being in it.

The philosophical approach (here also in the broader sense of the term) moves on a level of interpretation of reality that is distinct and different from those investigated by science in the modern sense. According to a classical understanding of philosophy, it begins with the unavoidable question raised by humanity about the final reason for the self and for all things, enquiring about the meaning (and final end) of what exists and of what happens. This approach presupposes that answers to this kind of question can somehow be found, as tentative and provisional as they may be, and that reasons, to be discovered and elucidated, exist and are given in the world. In this way, the philosophical approach to reality presupposes and expresses the profound perception of a threshold between what is here and now for me, spontaneously described and scientifically interpreted, and its principle and final end, certainly mysterious and beyond, yet rich with the promise of a truth capable of providing light and flavor to our existence and our destiny.

The theological approach is yet another and different approach. It also begins with a question about the reason for and meaning of things, but in this case, not as raised only by myself and based on the experience and tradition of thought that I have received and live amongst: rather it is raised—and this is the point—beginning with and in dialogue with God. In the religious experience, which has a specific realization in the experience of Christian faith, God himself (whom I experience, in wonder, from the heart, in the biblical sense of the term, that is, in the spiritual center of my existence) has taken the initiative to pronounce a word—for the Christian, *the* Word which, in order to speak to us in a definite and purposeful way, became man in Jesus Christ.

Fantasy, illusion, projection? It undoubtedly could be that, or could become that, but—based on the sincere experience of many and perhaps even my own personal experience—I must also take this specific approach to reality into account. I must do this even more so because it is not inexpressible, but, arising from a form of dialogue, albeit a completely singular dialogue with God, it is communicable through the fact that it tends to show in itself the reason (the *lógos*) that inhabits it. Moreover, because it is also a human approach, albeit enlightened by the light that comes from God, it therefore grows and develops, and can be made more precise, verified, and put to the test, and for this reason can, or perhaps better, *must* dialogue profitably with the other forms of knowledge. Even theology, therefore, like philosophy and science, can refer to a specific level in which reality is given and discussed.

Each of these approaches operates within its own field of exercise and offers something important, at its own level, regarding the questions of necessity and/or of freedom as keys for interpreting existence. The scientific approach, for example, in the work of Galileo, says something indispensable about the how the Solar System works, and in the work of Darwin, something crucial about the evolution of the living species. What is essential is that no approach should try to invade the other fields with its specific language, method, and goal, because in fact, and in principle, it would not have the competence to do so. Yet it is not always easy—in fact, quite the contrary—to interact with the others while remaining strictly faithful to one's own approach. In the end, the plurality of approaches tends to interpret reality, in a way that is coherent and beneficial for everyone, as something that is a whole in itself, even if it expresses itself on a series of different levels.

As previously mentioned, the three concepts that are fundamental in the interpretation of the sense of being from the theological point of view, which I have been called upon to illustrate and which are the starting point for my contribution, are precisely grace, freedom, and relation. I will say something about each of these in my attempt to set out a coherent description of the interpretation of reality offered by the theological, cosmic, and anthropocentric vision in the biblical-Christian Revelation.

If Truly "Everything Is Grace"

I begin with grace. I should say immediately that, in the terminology which has matured through great efforts over the centuries, from the heart of Christian experience and intelligence, the term does not merely refer to a sentimental and ultimately accessorial or even illusionary dimension of reality. According to the perspective proposed here, it involves a vision illuminated and inhabited by the light and the essence of truth. It is understood in the sense that grace speaks of the meaning and destiny of reality, insofar as it is originated, informed, and directed by gift and by forgiveness.

First of all, it is commonly said that reality "is given", in the sense that it is a gift which gives itself. Being a gift expresses the intimate and irreducible being of reality. Being given and being a gift set the rhythm from which emanates the perfume of the reality that it safeguards, not as accidental and contingent, but as substantial and abounding with the taste of the eternal. Being a gift, therefore, is an epiphany of the sense and truth of reality.

But that is not all, because grace also refers, with an intensifying determination, to forgiveness. As the meaning of reality, the word "gift" implies not only the intrinsic gratuitousness and excess of what is being offered, but also the gratuitousness and excess of its being recognized, welcomed, and offered again. Thus it is precisely the nature of grace as a gift to propose itself freely again and again in an excess of forgiveness, whenever the gift is not acknowledged or is misunderstood or even rejected. It is here— in this specific determination of forgiveness—that grace expresses the fact that it is a gift to the very end (*eis télos*, in the Scriptures).

In a word, using the expression of Georges Bernanos on the final page of his *Diary of a Country Priest*: "everything is grace!",[1] everything being illuminated in its essential truth by gift and forgiveness.

A window upon the mystery of grace as the original theological key to the interpretation of reality is opened in the experience led first by Israel and then by Jesus. Little by little and then rather suddenly an unexpected horizon

---

[1]Georges Bernanos, *Journal d'un curé de campagne* (Paris: Éditions Plon, 1936): «tout est grâce».

is revealed, in which the dialectic of destiny and freedom is rewritten from top to bottom. It is clear that this is not done in a pacified or pacifying way, but nonetheless in a form that is intense, enlightening, and provocative.

In the Old Testament, God speaks to humankind, saying: "I will safeguard you as the pupil of my eye" (Deut 32:10). The eye of the Divine, in this image, is not like the blindfolded eye of Tyche, the goddess of fortune, distributing good and evil, and deciding destinies by handing out good and bad luck. Rather, the eye of God looks upon humankind with a view to protecting them as what is most precious and intimate to God himself: just like when someone blinks to protect their pupil from being harmed by the sun or by a piece of flying dust.

Yet what does it mean, and what is behind such an experience of feeling and knowing that one is being watched and protected? What kind of grace are we faced with in this view? And what happens to the destiny and freedom of humankind? Our thoughts go straight to the apostle Paul and to the remarkable text *Letter to the Romans*, which has drawn the attention of so many throughout history, and rightly so: from Augustine to Luther to Karl Barth. It is in this letter, in fact, that the manifesto of grace is indelibly written, starting with the events in which Jesus the Christ was crucified, then rose again. However, to reach an understanding of its meaning, we must first of all provide a quick overview of what precedes this dramatic moment when Paul experienced grace in Christ, as attested in his letter.

Let us return, therefore, at least for a moment, to the verse of the psalm previously mentioned. It all begins with the fact that ancient Israel experiences the benevolent (and demanding) gaze of the Lord God. The moment the Lord hears Israel cry out to him in pain from the land of Egypt, he "comes down" to free the Israelites. There was certainly a free and unmotivated choice at the origin of this, an election made by the Lord for his people that would be sealed with a pact at Mount Horeb. The book of *Deuteronomy* describes this choice (4:32–40). It was a free act of election, but at the same time it was a demonstrative act to awaken the people and lead them towards a shared awareness, beyond ethnic and religious borders, that everyone was looked upon by this gaze. This is how Psalm 33 expresses it: «From his dwelling place he watches all who live on Earth, // he who forms the hearts of all, who considers everything they do».

The Bible prefers the Hebrew word *chen* to refer to this particular attitude of God towards humankind, which the Greek version of the LXX translates for the most part as *cháris*, grace. Two meanings are conveyed by the Hebrew term: first and foremost, benevolence, in the originating sense of wanting what is good for others and looking upon others with kindness and without envy; and secondly, mercy, in the sense of having a tender heart that knows how to understand and forgive.

Many Hebrew words are used to express the experience of divine mercy and benevolence, but there is one that is especially suggestive and rich in meaning: *rahamim*. It comes from the root word for womb, uterus: *rehem*. Thus *rahamim* has a feminine and maternal connotation alluding to the visceral relationship that a mother has with the fruit of her womb. It is the translation of this term that we hear in the liturgy as the "bowels of mercy" (*viscera misericordiae*). It refers both to the attitude of benevolence and forgiveness and to its ultimate reason and root: the visceral love of a mother for her own son. Having made this point, we can now focus on two important considerations.

(a) The Jewish perception and semantics of grace in itself contains an original and almost inextricable polarity, at least at first glance: because mercy is different than benevolence. In fact, benevolence means wanting the good for others and in itself connotes the identity and action of God whose name is the Lord: that is, "I am and will be with you" (see Ex 3:14). He, therefore, wills the good and only the good of others. This is what resonates most clearly and majestically in the first page of the Bible, in the account of the creation. The "let there be light" (Gen 1:3) pronounced by God, with all that follows, is a free, gratuitous, benevolent act that finds confirmation in the statement God makes about the effect of his action: "and God saw that it was good" (Gen 1:4), an affirmation that expresses both wonder and satisfaction and, in the case of the creation of man and woman, becomes: "and God saw that it was very good" (Gen 1:31).

The mercy of God, on the other hand, shows his obstinate will to go beyond the hesitant and imperfect way with which humanity responds to his benevolence, even offering, of his own initiative, to re-establish the relationship when it has been interrupted or betrayed or refused. In other words, mercy intensifies the gratuitous and relational intentionality of benevolence. Forgiveness reveals the free and unlimited abundance of the gift. Moreover, in and through forgiveness, God gives even more than was promised and given. Mercy, therefore, makes tangible, on the side of God, the excess that God promised in benevolence, and, on the side of humanity, the measure of responsive and responsible freedom, as implied and encouraged in relation to mercy.

(b) The second consideration involves another formidable antinomy that the Jewish experience and semantics of grace exhibit and inspire: the one established between the particular dimension of election and the universal dimension of the gaze of God, which is expressed, for example, in the account of the creation in Genesis. Produced by the event of grace itself as embraced by Israel, this antinomy necessarily causes acute tensions. Yet it is progressively perceived as insurmountable, in that the two poles of the particular and the universal dimensions of grace are to be held together, come what may, so as not to betray in a destructive way the novelty of God's irruption into human experience and history. In the book of *Deuteronomy*, for example, the idea of election is developed as the nonnegotiable principle which grace depends upon, as shown by God towards Israel. Yet the prophets—from Amos to Jeremiah—do not like to speak of election, for fear that it be understood as an automatic guarantee of salvation, closing an exclusive circle that implies the exclusion of others. The idea of election, in the experienced awareness of the theological aporie it conceals, is therefore balanced, on the one hand, by the idea of a possible "rejection" by God (see Jer 14:19) and, on the other, by the intrinsic reference of the election to a universal project that is destined to call upon all peoples (see Psalm 87).

The theology of the apostle Paul surely accounts for this rich, though internally even antinomic, inheritance. The concept of *cháris* is absent in the synoptic gospels, with the exception of some occurrences in Luke, while in the gospel of John it is present only in the Prologue (1:14–17). In Paul, however, it is definitely a central theme, since it best expresses the meaning and dynamics of the salvation event that God so freely and paradoxically produced in Jesus Christ for the benefit of humankind.

The heated nucleus of the Pauline doctrine of grace flows forth from the fact that, in Jesus, the definitive and irrevocable "yes" of God's love—*agápe*—resounded in the world. For Paul, this is the grace of Christ, the grace that is Christ himself, Christ present and operating in believers through his Spirit. This is the conclusion reached by Paul after encountering the risen Jesus. In Christ's story, Paul finds the key for reading the plan that was hidden for centuries in the foreknowledge of God and finally realized in the fullness of time. This is the *focal point* from which Paul looks upon everything and demonstrates his discourse about grace with tenacity, passion, and impetuosity: because he judges it to be decisive in the proclamation of the gospel of Jesus Christ. In the concise argumentation of the *Letter to the Romans*, in fact, the interpretation of the *cháris* of God in Christ allows Paul to propose in a new way the two antinomies which, as we saw, connote the experience and understanding of grace in the First Testament: the one between gift and for-giveness, and the one between particularity and universality. He does so, not to seriously reduce the two poles of these unavoidable tensions, but rather to show their intrinsic dynamics and effectiveness.

(a) Let us begin with the second antinomy, as Paul himself does. Jesus Christ represents, for him, that singular event of grace *from God* which is made possible by the particularity of the election in relation to Israel. Upon the wood of the cross, it is opened from within itself to the universality of all peoples. This is because, in Jesus Christ, the unequivocal offer of God's grace is witnessed and shown to everyone, Jews and pagans alike. No one can claim privilege or merit. God's initiative is absolute, gratuitous, and universal. Therefore, it is not belonging to the people of Israel, nor performing the works associated with the observance of the Law given to Moses that justify one before God.

As Paul exclaims: «Does God belong to Jews alone? Does he not belong to Gentiles, too? Yes, also to Gentiles, for God is one and will justify the circumcised on the basis of faith and the uncircumcised through faith» (Rom 3:29–30). This is central, the gospel of grace: «There is no distinction—Paul insists –; all have sinned and are deprived of the glory of God. They are justified freely (*doreàn*, by pure gift) by his grace through the redemption in Christ Jesus» (Rom 3:22b–24). Faith is unconditional openness to *this* grace. It "justifies", meaning it renders us just before God, because it is the acceptance of God's gift and for-giveness in Jesus Christ. It is God, therefore, who by grace makes us righteous, meaning new and capable of walking henceforth in justice, in conformity with the grace received and embraced.

It is *by love* that grace justifies and frees us from sin (which is closure within oneself, in relation to God and others, to the point of implosion). In fact, grace is nothing other than the overwhelming attestation, in Jesus Christ, that God is *Abbà*, Father, and that we are sons. The reality, awareness, and exercise of that is precisely grace, meaning a free gift, not only in the sense that they are objectively bestowed *upon us* by and in Jesus Christ, but also in the sense that their acceptance *in us* is the fruit of the Holy Spirit, that is, of the presence of God's love itself as the breath of life in *our* freedom. As Paul explains: «For you did not receive a spirit of slavery to fall back into fear, but you received a spirit of adoption, through which we cry, "Abbà, Father!" The Spirit itself bears witness with our spirit that we are children of God» (Rom 8:15–16). In this logic, grace is the principle of freedom: it enkindles, promotes, and requires it.

(b) The other antinomy contained in grace, the one between gift and forgiveness, also receives new light from this *focus*. In fact, it is clear that the first obvious product of grace is forgiveness. Yet the experience of forgiveness is none other than the wide open door to receiving the abyssal gratuitousness of God, upon which all things depend. By being and acting in the regimen of gift, brought to its highest expression in the gift of self, God constitutes others in their capacity also to be themselves through self-giving. Grace, therefore, reveals, for Paul, the astonishing law of excess and abundance that regulates God's being and actions with its measure beyond measure of freedom and love. This same law is called upon to regulate the being and action of humankind, as His image and likeness.

Paul's theological intuition here allows for further understanding of the election, which is also gauged from top to bottom by the experience and understanding of grace in Jesus Christ. It is not the bestowing of grace which is commensurate with the election that predetermines its quality and recipients; rather, it is the election which is commensurate with the measure of grace beyond measure that occurs in Jesus Christ. This principle, derived from the salvific event of Jesus Christ, is evident in the *Letter to the Romans*, especially where Paul speaks, in Chapter 8, of «those who are called to his purpose. For those he foreknew he also predestined to be conformed to the image of his Son, so that he might be the firstborn among many brothers» (Rom 8:28–29).

To be pre-established or predestined, acknowledged for those who love God, is not to be understood in the sense of a separation from those whom God may not have pre-established or predestined—an interpretation that would prove tempting to some and gain importance in the later theological tradition. No. As Heinrich Schlier points out,[2] those who are called qualify as predestined "so that it is clear how God precedes those who love him". In fact, "he has predestined humankind from the beginning—and that is clear in those who love God, who have answered God's call—to become sharers in the being of Christ".

The Father's election is «before the creation of the world» (Eph 1:4), and therefore radically precedes any consideration of human responsibility in history, be it good or bad. Yet it passes through the redemption of all in «the blood of Christ» (Eph 1:7). In him, the grace of God is conceded "at a high price"—to use the words of Dietrich Bonhoeffer –, without holding anything back on the part of God.[3] Humankind is therefore asked to receive it with an acceptance that is unarmed, of course, yet serious, active, and responsible. Grace is not deserved through one's deeds; it becomes operative in faith through love (see Gal 5:6).

The grace that reaches us and that we are called upon to embrace and live is not chance or destiny, dispensed with eyes closed by "fortune", but the gift willed by God's love for all and entrusted to each person's freedom. Certainly, the inevitable antinomies of grace expand the horizons of our freedom and love to infinity. So to accept grace in our existence and in our intelligence requires something radical and paradoxical, which is signified by the cross of Christ. This is how Simone Weil describes it, in her unique and striking way: «Grace fills empty spaces, but it

---

[2]Heinrich Schlier, *Grundzüge einer paulinischen Theologie* (Freiburg im Breisgau—Basel—Wien: Herder, 1978).
[3]Dietrich Bonhoeffer, *The Cost of Discipleship*, Trans. by R.H. Fuller (New York: Macmillan, York 1963).

can only enter where there is a void to receive it; and it is grace itself which makes this void».[4]

Freedom, the Only True Place for the Encounter Between God and Humankind

To say this about grace—or rather, to have this experience and, from within it, investigate our understanding of reality—means to say that "everything is freedom" or—as Luigi Pareyson[5] liked to say—that there is only one thing that I am not free to do and that is not to be free!

However, this does not mean that all reality, at its different levels of realization and in the different kinds of interpretation that they require and propagate, is to be understood according to the terms of freedom that are attributed in a personal way to God and to humankind; nor does it mean, on the other hand, that freedom is to be understood in absolute and arbitrary terms, as something that is completely unrelated to anything else. The fact is, rather, that the experience of freedom, and the corresponding insight of intelligence to specify its meaning, which unfold from the horizon of truth we call grace, can be precisely delineated. The freedom we are dealing with here is not just self-determination as the possibility of choice, but the concise expression of the human being. This expression occurs both in freedom being guaranteed and founded by and in the grace of God, or better, by God who is the gift of self and therefore himself freedom,

---

[4]Simone Weil, *La Pesanteur et la Grâce* (Paris: Librairie Plon, 1948). Own translation.

[5]Luigi Pareyson, *Ontologia della libertà. Il male e la sofferenza* (Torino: Einaudi, 1995).

and in its full realization, precisely as graceful freedom which in turn expresses itself in the gift of self.

In other words, the experience and understanding of grace simultaneously require and propagate the experience and understanding of freedom, both of God and of humankind. If, in fact, the key to interpreting the sense of reality is grace, freedom results as both its condition of possibility and its effective realization. Without freedom there is no grace, just as without grace there is no freedom. It is no accident that, among the first Christian theologians, in the second century after Christ, Irenaeus of Lyon, filled with the Spirit that springs forth from the New Testament witness of the event of Jesus Christ, not only emphasizes the fact that Jesus's is the "gospel of freedom", but he dares to make the following statement: «He [the Creator] made all things freely, and by His own power, and arranged and finished them, and His freedom is the substance from which He drew all things».[6] This statement, if we situate it in the context of his thought, is to be understood in the sense of the ontological correspondence between the freedom inscribed in the reality of God and the freedom of humankind, and through humankind, of the cosmos.

Within the horizon of sense and truth disclosed by grace, we may say that freedom is the only true place of encounter between God and humankind. If the human being were not to access God through liberty and as liberty, it would contradict his being human: and not only that, but it would also contradict God himself in his most intimate and mysterious being. Moreover, we can definitely say, looking at the history humankind has shared with God, that the wearisome and often tragic experience of human freedom goes hand in hand with human

---

[6] *Adv. Haer.*, II, 30,9; see IV, 20, 2.

experience of God's freedom. This is so clearly the case that, throughout the upheavals human history, one is not given and does not occur without or against the other.

«Eàn oûn ho huiòs humâs eleutheróse, óntos eleútheroi ésesthe»—according to the fourth gospel (8,36): «if therefore the Son frees you, then you will truly be free». Freedom is the grace of the Son: the Son of man who is the Son of God and who—in his Passover of abandonment and resurrection—becomes the epiphany of the freedom of God and the epiphany, literally in the specific place of being (expressed by the Greek adverb *óntos*), of human freedom. This is the crucial task entrusted by his Spirit to our history.

It is difficult for us to conceive and rejoice in the measure of freedom that was unexpectedly and joyously unveiled for the disciples in their encounter, first with Jesus of Nazareth, and then with him as risen. Yet upon a closer look, this is the culmination and fruit—albeit surprising and unforeseen—of a path toward freedom that was opened at the very outset of the history of humanity, thanks to God having constantly made himself present and thanks to his unpredictable irruptions in history, which little by little created new thresholds of consciousness and responsibility. The torment that has afflicted human history since primordial times can in fact be summed up in these terms: how is human freedom possible within the sphere we live in, which is closed upon itself? Is not humanity, along with all the gods, subject to the inevitable trajectory of destiny, which gathers all together and guards them? After all, this is the perception that often ran through human experience and thought; the impenetrable veil of fate seems to surround the existence of the cosmos, and it is only from within the dense net of its web that a limited though responsible measure of freedom is permitted. It is due to this measure of freedom that humankind is called to serve justice and virtue.

As Plato wrote in reference to the myth of Er in *Politeía*, «Virtue has no master»,[7] and Cicero identifies the specific quality of *civis romanus* as being servants to the law in order to become free: *«legum servi... ut liberi esse possimus».*[8] Yet, for Plato, the choice of one's own life project depended upon what happened in the previous life and remained shrouded in the oblivion of time immemorial.[9] While for Diogenes Laërtius, the wise and the just are free only because their actions conform as closely as possible to the needs of the cosmic and social order.[10] In this context, perhaps the highest intuition of the only lever that can disconnect the world from the wheel of destiny was offered by Buddha. As in the case of all the pearls of truth and the fruits of justice that have matured throughout the history of humankind, across all latitudes of religion and culture, such an intuition is no stranger to the discreet but effective stimulus of the light that comes from God. In fact, in the abyssal depths of his interiority, Buddha intuitively experienced that the sense of freedom comes from beyond the world and that it can be obtained only in a nullification of the world itself, with its inevitable chain of causality. Thus, by exercising universal compassion, the dawn and power of true freedom can at last shine forth from this empty nothingness.

The experience of Israel fits in here, though along a different path, which ultimately seeks the answer to the same yearning. Israel's path is one in which God's invitation is not to leave history behind—not even to come back to free it from the bonds of fate—but rather, He becomes a companion of humankind in history, coming down

---

[7] *Politeía*, X 614A–621D.

[8] *Pro Cluentio*, 53, 146.

[9] *Politeía*, 620A.

[10] VII, 88; see Cicero, *De fato*, 17.

himself to "free Israel from the hand of Egypt" (as in Ex 3:8; 20:2; see also Deut 26:7–9). The freedom pursued by God is the one that decides the life of humankind in history. It seeks to create the specific conditions and place where human beings can express their freedom vigorously and authentically among themselves. The experience of freedom is both a gift and a responsibility. But first of all and ever renewed, it is a gift that spreads and promotes the exercise of responsibility. In the dramatic history of Israel, therefore, freedom takes on worldly and communitarian connotations. These are the grounds upon which the profile and identity of men and women take their shape: created "in the image and likeness" of God himself (Gen 1:26), they are free, in their mutual relationship, which opens upon the world. The experience that slowly becomes consolidated, but not without divisions, failings, and fresh starts, is that the Lord, and only the Lord, exacts and guarantees the freedom of humankind. In truth, He himself is the freedom of humankind. It is in relation to Him that human beings acquire the freedom to be themselves. This explains the eternal and critical struggle against every form of idolatry. He is their freedom also in the sense that, as His divine will is transplanted from His heart to theirs, the amazing amount of freedom that is in His heart can also germinate in the heart of humankind. This is what the Lord promised through the voice of the prophet: «I will give you a new heart, and a new spirit I will put within you; and I will remove the heart of stone from your body and give you a heart of flesh. I will put my spirit within you so that you walk in my statutes, observe my ordinances, and keep them» (Ez 36:26–27).

Nonetheless, Israel's experience of freedom is not without contradictions. First of all, the one that occurs as the law of freedom is petrified into a law of slavery by the human heart of stone. This contradiction is stigmatized by Jesus and by Paul as the prerogative, so to speak, not only of Israel, but also of any faith experience in a covenant relationship with the personal Lord God. In their experience of freedom, Israel is ruled by a God who is the sovereign and invincible Lord of freedom. His freedom is measured only by His faithfulness to the promise and grace towards humanity that He himself unquestionably decreed, which is therefore God's fidelity to Himself. Magnificently showing how God is free only and always in faithfulness to Himself, the Bible provides a chink of light so dazzling it is dark, and thus so difficult to grasp in its gratuitous unfolding in history.

Certainly, as we saw in what Irenaus of Lyon said, the gospel of Jesus Christ is essentially the "gospel of freedom", as the gospel announced by Jesus and the gospel that is Jesus himself. The tortured path of freedom seems to lead towards Him: even if that can only be said—following the logic of Christian faith—after the fact and, once again, not without experiencing the acute laceration of contradiction. It is impossible, albeit truly fascinating, to perceive here how the freedom of God and the freedom of humankind are realized before our own eyes in the figure, *kerygma*, and actions of Jesus of Nazareth. What we *can* do, however, is try to identify the source of his freedom. How can Jesus radiate and disseminate freedom?

Jesus lives off freedom, *because he is freedom*. His freedom coincides with his adult relationship as Son before the *Abbà*. The Breath of freedom abides in Him, and is spread to his surroundings by Him. It is born from and forged in his relationship with the *Abbà*. The numerous threads of the history of freedom between God, Israel, and all peoples are surprisingly and unexpectedly tied together in this relationship. It is the experience of God/*Abbà* that Jesus has and is which determines the freedom he bears witness to in his proclamation, his life, and his death. His freedom expresses his being the Son, that is, his being before God as One who measures and gives his own life in the hope that others may exist and live, as a gift, what He himself lives. In this sense, God/*Abbà* is truly and fully the source of Jesus' freedom. This is not to be taken for granted: because if freedom was a gift of the Father, it was at the same time—in the history of the Son who «became flesh» (Jn 1:14)—a painful achievement, an agonizing decision, and an endless risk. Episodes of struggle and suspense in great distress testify to this: in the temptations in the desert at the beginning of Jesus' ministry (Mc 1:12–13; Mt 4:1–11; Lk 4:1–13); in the mortal anguish he suffered in the olive garden towards the end of his ministry (Mc 14:32–42; Mt 26:36–46; Lk 22:40–46); and even more so in the cry of abandonment from the cross (Mc 15:34; Mt 27:46).

Here the Father's gift of freedom to the Son becomes one of the greatest dramas in the history of humanity. The freedom of God that came down from above died in the furrows of history, if we can borrow the metaphor from the Bible, so that it could germinate in the life of humankind: «unless a grain of wheat falls to the ground and dies, it remains just a grain of wheat; but if it dies, it produces much fruit» (Jn 12:24). The freedom of God, and God alone, is faithfulness to Himself in his will for the other to exist and to have life in himself (and this is the Son, Jesus). The immensity of this freedom, which is specific to God, is poured out in the same extreme measure of freedom associated with Jesus, the Son, who freely and unconditionally believes—without the need for any form of reassurance—in his Father's love. He continues to do so even when the Father is silent and does not intervene to defend the cause of his Messiah, and even when everything around him seems to be proclaiming the very opposite of love, the root and fruit of freedom. Allow me to make a twofold observation about this.

(a) On one hand, the New Testament registers the dramatic tension between Jesus' freedom and the will of God, *Abbà*, in the noted episode of Gethsemane. Without any attempt to soften the effect, it shows Jesus' mortal anguish in adhering to the will of the Father; and it thus shows the extreme and risky nature of this freedom, called to conform, not to blind necessity, but to «a costly grace» (as Bonhoeffer says), referring to the gift and forgiveness of God as the alpha and omega of the meaning and destiny of truth. His anguish expresses the extreme challenge involved in embracing his freedom, that is, in conforming to the freedom of God, which is entirely expressed through His gift and forgiveness.

(b) On the other hand, the same New Testament refers to what Jesus said at this point, which sheds light upon the meaning of the above drama embodied in his passion. His freedom (*eleutería*) is shown to be the substantial expression of his *ex-ousía* (literally, what comes from the substance of his very being): «This is why the Father loves me, because I lay down my life in order to take it up again. No one takes it from me, but I lay it down on my own. I have power (*exousía*) to lay it down, and power (*exousía*) to take it up again» (Jn 10:17–18).

This is where it becomes clear that Jesus exercises true freedom, because it is realized in accordance with the Father's freedom, in the gift of self that is also forgiveness, recapitulating in itself the destiny of all reality. His belief in the love of God/*Abbà*, dedicated to his brethren to the end (*eis telos*, Jn 13:1), in response to the freedom of God «who first loved us» (1Jn 4:19), has now been definitively (*ephápax*) rooted in the history of humankind: «On Earth as it is in heaven.» Jesus opened a window upon the staggering abyss of God's freedom and its traits, such as the *agape* that gives his life so that the other may exist and be like himself, in the grateful acknowledgement of the gift received (of the gift he himself is). From the very heart of history, God's freedom is thus offered in Jesus as a measure of the grace and truth of human freedom with and for others before God. Yet this measure is not the sole property of Christians, but rather, it is the definitive and irrevocable inheritance of all humankind. Nonetheless, the temptations, which Christ had already overcome in the desert and throughout his earthly existence, need to be avoided from now on, so that in the Spirit of Jesus, God's freedom may germinate from Heaven and prosper as human freedom on Earth.

Indeed, the question of freedom has been challenging human conscience, thought, and action for several centuries now. Not that it had not done so before, but there is no doubt that modern times bear particular witness to the unavoidable appearance of this critical challenge on the scene of history. It has been such a question that today, having reached, willing or not, the terminus of modernity, we are also bluntly forced to realize that, not only can freedom be expressed in many different ways, but its understanding and exercise have entered a new phase in its history. Certainly, freedom should be approached in rigorous and open dialogue with neuroscience, philosophy, and theology. Such an approach creates a space in which, and through which, the different kinds of access to freedom through the relevant disciplines can interact with each other, while respecting the specific characteristics and autonomy of each, and at the same time being dialogically articulated according to a overall polychromatic and coherent plan. The methodological principle of "distinguishing to unite",[11] or in other words,

---

[11]Jacques Maritain, *Distinguish to Unite, or, The Degrees of Knowledge*, Trans. by Gerald B. Phelan, University of Notre Dame Press, Notre Dame (IN) 1995.

reflecting and acting "without separation and without confusion", once again bears fruit.[12]

At the same time, such an approach must be open to the diachronic dimension of the history of freedom, which is also connected to the history of the world in which we live, and which we ourselves constitute. Indeed, the passage from neuroscience and philosophy to theology invites us to reread the *quaestio de libertate*, not only distinguishing between its various levels of expression (biological, anthropological, theological), but also looking at the way each of these emerges in the specific history of the development of the created universe. In this context, we may examine, at least from a phenomenological point of view, the impact that the event of Jesus Christ had on history in the tremendous drive for a self-awareness and self-configuration of freedom on the human and social levels. The Christian tradition bears witness to this in its most specific meaning, and so does the complex and ambiguous history of modernity.

Yet there is more to it. The impact of Jesus Christ upon the history of humankind (and the world) has brought about a true change of perspective. Of course, this encourages us to look at freedom "from below", in the biological and anthropological development of its conditions of possibility and the specific way it is exercised socially, even though—as Kant taught us—its effective possibility derives from another order.[13] Yet at the same time, it also encourages us to look at freedom "from above", that is, by identifying the logic of God's plan for creation as it unfolds in Jesus Christ, who corresponds—in human

---

[12]According to the article of faith defined by the Council of Chalcedon (451); Heinrich Denzinger, *Compendium of Creeds, Definitions, and Declarations on Matters of Faith and Morals*, Edited by Peter Hünermann for the original bilingual edition and edited by Robert Fastiggi and Anne Englund Nash for the forty-third English Edition, San Francisco: Ignatius Press, 2012, no. 302.

[13]Immanuel Kant, *Kritik der praktischen Vernunft* (Leipzig: P. Reclam, 1878).

form—to the very event of God. Being *Agápe*, in the reciprocal and open gift of Self to Himself in Himself, God is a Trinity of persons, as testified by Christian faith. It is clear that this logic is specifically theo-logical. However, because of its intrinsic nature, it does not diminish the value and significance of the achievements made along the way by bottom–up thinking. If anything, it does justice to them as consequences of a specific approach, recognizing the fact that they are ultimately founded upon the logic of the historical unfolding of God's (Trinitarian) plan for creation.

In Jesus, the revelation of the co-original dignity in God of identity and otherness opens the theoretical space for creation as it unfolds *from* God, according to the plan for its gratuitous and free fulfillment *in* God. Creation thus becomes fully itself in the specific form of relation *with* God, who establishes it in its otherness. Such an ontological hermeneutic in Trinitarian terms expresses the inherent meaning of the freedom of being, in God and in humankind.[14] This is to be delineated in at least two steps. The first involves developing the explicitly ontological potentiality of the category of possibility. By expressing the self-determination of freedom, which assumes the intentionality of what is and what should be, possibility is achieved for what it is only in the relation of the self with others, as the self is implicated in the very exercise of possibility. The category of possibility, therefore, interpreted according to its ontological significance in the personological and ultimately Trinitarian perspective, is what it is only if exercised and intrinsically understood in the

---

[14]Klaus Hemmerle, *Thesen zu einer trinitarischen Ontologie* (Freiburg: Johannes Verlag, 1992). Allow me to refer to my *Dalla Trinità. L'avvento di Dio tra storia e profezia* (Rome: Città Nuova, 2011), the English translation, *From the Trinity*, is being published by the Catholic University of America Press, Washington DC.

dynamic of com-possibility. We thus find ourselves faced with the Rubicon which—as von Balthasar would say—metaphysics has to cross, since it arises from an existence of being and thinking in Christ: the threshold that leads from the individualism of the substance to the interpersonality of the relation.[15]

Crossing this Rubicon opens the way to a second step that must be taken in order to explore the potentiality (of experience and intelligence) made available by going in this direction. In the delineation, so to speak, of its transcendental conditions, freedom is shown to be not only compossible, but also effusive; or better still, it reciprocates reciprocity *in actu*—that is, ontologically.

Relation as the Truth of Being

In the logic of what has been discussed so far, we have already touched upon the third term of the triptych proposed at the beginning: relation. If, in fact, grace is given as and in freedom, then freedom is given as and in relation. So much so that, from a theological perspective, an equation can be proposed between the ontologies of grace, freedom, and relation.

In the metaphysical framework proposed by Aristotle, relation is merely an "accident" of substance, and moreover the least, the most fragile, and accessorial among the "accidents" of substance.[16] Yet for Augustine, a brilliant interpreter of the *novum* introduced in the vision of being by the biblical Christian experience, the relation *in*

---

[15]Hans Urs von Balthasar, *Theo-drama: theological dramatic theory*, 5 Vols. (San Francisco: Ignatius Press, 1988–1998).

[16]See Aristotle, *Categories*; http://classics.mit.edu/Aristotle/categories.1.1.html; Id., *Metaphysics*; http://classics.mit.edu/Aristotle/metaphysics.html.

divinis expresses the very meaning of substance, that is, of being itself, in what makes it what it is, in the expression of its meaning.[17] Along the same lines, in the Middle Ages, Thomas Aquinas forged a definition of God's very being—*Ipsum Esse per Se subsistens*—as *relatio subsistens*,[18] according to which it is given only in the ever renewed and boundless relation of reciprocity between the three divine persons. This line of thought has been brought up to date by Antonio Rosmini, according to whom every personal being—not only God but also human beings— is in itself a relation.[19] And this is so, insofar as being is freedom, and freedom is an "I" which becomes itself in the gift of self to another, thereby giving life to the "we" of open reciprocity.

Is this not confirmed by our own experience? We say that we feel "free", not so much in the self-determination of ourselves, as in the successful establishment of a relationship with others: "with you, in this situation, in this relationship, I really feel free?" Freedom is achieved as such in the context of grace, when we are given what is in itself free, that is, where freedom is given in and as relation. And not just any relation, but that of reciprocal acknowledgement. Ultimately, the relation—as I like to say—is that of "reciprocating" reciprocity, so it is a relation that is not closed and exclusive, but open and boundless, propagating to infinity.

---

[17]Augustine, *De Trinitate*, V, 5.6: «*Quamobrem quamvis diversum sit Patrem esse et Filium esse, non est tamen diversa substantia, quia hoc non secundum substantiam dicuntur, sed secundum relativum; quod tamen relativum non est accidens quia non est mutabile*».

[18]Thomas Aquinas, *Summa Theologiae*, Ia, q. 29, a. 4; cfr. A. Krempel, *La doctrine de la relation chez Saint Thomas. Exposé historique et systématique* (Paris: Vrin, 1952).

[19]Antonio Rosmini, *Theosophy*, Trans. by Denis Cleary and Terence Watson, 3 Vols., Rosmini House, Durham 2007–2011, n. 903.

A mystic of our time, Chiara Lubich, referring to the central and generative truth of the Christian vision of God as a Trinity, concluded that «*the Trinity is freedom*», meaning that freedom is given within the space of reality described by an infinite reciprocating reciprocity.[20] In light of this, if we consider the Christian event for what it is from a theological point of view—as the truth (of God) offering itself in the history (of humankind)—then it becomes possible and necessary to give word and reason (*lógos*) to the Christian event as constitutive of and thus revealing the sense of being. This implies the exercise of a *lógos* that not only respects the identity and the vocation of being thus constituted, but also itself reveals and promotes being. It is precisely to this intense center of speculation that the *lógos* is called to give reason to the existence we experience, or as Luigi Pareyson would say, to explain how the relation with oneself coincides with the relation with others.[21]

This is not just about forging the classical concept of substance as what is conceived in and through itself; nor the modern concept of subject, as what is immediately or mediately transparent to itself. Rather, it involves thinking about existence in the truth that is given historically of its being relation to itself while being relation to another. This coincidence between "self-relation" and "hetero-relation" in the identity of existence thus expresses, in the most concise and precise way possible, the epochal turning point suggested to the responsibility of the *lógos* after modern times. The task at hand is to go back to the timid initial *lógos* of Augustine, renewed and superbly explored—but only *in divinis*—by Thomas Aquinas in the Middle Ages,

---

[20]Chiara Lubich, *Essential Writings: Spirituality, Dialogue, Culture* (Hyde Park, NY: New City Press, 2007).

[21]Luigi Pareyson, *Esistenza e persona* (Genova: Il Nuovo Melangelo, 2002).

and which has come back on the scene in the dissolution of modern times, as intuited by Rosmini. This reconnects with the beginning, but from within modernity, therefore shifting from the theological to the anthropological and cosmological horizons, and thereby proposing a reformulation of them. The relation *with* God, or better still, *of* God who is the foundation of the freedom of the individual, becomes in itself the foundation of the interpersonal relation. The *lógos* is called upon to reflecting upon this and the existence in the historical situation of its being the relation from and to God.

The coincidence between "self-relation" and "hetero-relation" that dwells within our experience, in the depths of a void which can never be filled, is rooted in the absolute coincidence with relatedness that is specific to God. Pareyson describes how God is «absolute irrelativity, yet He poses a relation». This is the truth consigned to the *lógos* by the being which opens itself to the Christian event. Opening the self to itself, by being open to the Christian event, is the actuating revelation of the same gift and the same task. God has to be God, who is defined in relation to himself alone: so that existence can recognize him as such, recognizing him as its ultimate truth. For this reason God is to be acknowledged as the principle and the goal of a relation which must *first* be real and true for God, in order to be real and true for existence. Since God is the coincidence of absoluteness and relatedness, following Pareyson once again, it cannot be said that relation includes God, but that God includes the relation, since He is such a term of relationship that He is at the same time the condition of relationship. Of course, the relation is defined in different ways according to how it is viewed, moving from God in reference to humankind or from humankind to God. Nonetheless, insofar as it is real and true because it is postulated (by God) and accepted (by humankind), the relation is given in the same form: as freedom. Thus from the heart of the ontology which the *lógos* is called upon to understand and give voice to, we are brought to the task of formulating an ontology of liberty as the expression of an ontology of grace and constitutive and revelatory of a radical ontology of the relation. Proceeding in order, let us first ask in what sense relation is to be considered *from* God as freedom.

This must obviously occur without contradicting the absoluteness of God as God, and the otherness of humankind as the receiving end posited by the relationship *of* God with humankind, and therefore as received. The relation must thus be taken, in reference to God, as the sign and fruit of a gratuitous excess that gives itself, and in so doing constitutes the other term by entering into the relationship. The gift of self is the foundation of the relation. Being is itself revealed in such a relationship, which can only be expressed and thought of as a gift. The sense of "gift" here is not merely *to give* something (but to whom, if there is no recipient?), but *self-giving*, that is, the giving of self which, to be real and true, presupposes the establishment—through gift—of the other as other, inasmuch as the other is able to receive himself by receiving in freedom the gratuitous self-giving of the one who was at the source of the gift. Hence, in asymmetric reciprocity, the *lógos* is called upon to express the relation from the receiving end that is humankind. The human being is in fact in a free relationship with God because the human being is posited as such by God. Relation exists, therefore, in the dynamic coincidence between "receptivity" and "activity". By receptivity, the person is ontologically constituted at the receiving end of a relation which calls for recognition and is to be exercised as such. Activity involves living the relation, given and received, as relation, and thus in freedom. In this way, as concluded by Pareyson, the "passivity" that reveals the asymmetric situation of the relation—in that it is freely offered by God—«is nothing but the diaphragm between two activities, one of which takes place upon the extension of the other», such that the freedom of existence is revealed as the «initiated initiative and consent to a gift».

More can be said, continuing the same line of thought. Indeed, how can human freedom manifest itself in existence, effectuated in response to the "extension" of the freedom of God? It happens every time someone gives himself to God by giving himself to another: «whoever does not love a brother whom he has seen cannot love God whom he has not seen» (1Jn 4:20b). The gift of self, in which freedom is realized to its fullest measure, is certainly the response to God's gift of self that establishes the human being as freedom: because God's gift of self is precisely the ontological foundation of the other as other than himself, and that is, as freedom. But *how* can this freedom (of humankind) receive itself as taking place upon the "extension" of the freedom that is God, from whom it originates, if not by being given in turn? This occurs within, not outside, the historical situation of humankind, that is, in relation to another who is also the receiving end, historically situated, of the relationship in freedom from and to the Other. Is this not the direction in which the *lógos* who «became flesh» (Jn 1:14), the anthropic *lógos* of the cross (1Cor 1:18), invites us to look? In his incarnation and crucifixion in the tragic reality of history, he expresses and promises the "self-relation" that coincides with the "hetero-relation". This occurs within human existence, of course, but it occurs first of all in God himself, in the depths of his very Being. The coincidence between "self-relation" and "hetero-relation" does not occur only in the relationship that he establishes in freedom as the source and end in relation to existence. It takes place specifically within Himself. What the *lógos*, in light of the event of Jesus Christ, is called upon to think about truth, with existence as the starting point, is in fact the otherness *in* God (the *Abbà* and the Son in the Spirit) as the revelation of the liberty *of* God and, for that reason, the space for the reality and truth of being *other than* God. As such, it is real and true—upon the "extension" of the gift of self *in* God and *from* God—in the gift of self *between* human beings: «love one another as I have loved you» (Jn 13:34). This is the task and promise of the revelation of freedom brought about by grace which is relation, therefore, in the *lógos* of being that finds its measure in the *lógos* of Christ crucified.

Conclusion: Return to the Premise

Clearly, the theological understanding of the biblical Christian inheritance that I have proposed can and must interact with a philosophical understanding. In conclusion, we should ask whether it can and must also interact with the understanding offered by scientific rationality, in the modern sense of the term? I think that the answer, albeit challenging and arduous in its elaboration, can only be positive in this case.

It suffices to recall, for a suggestive and thought-provoking example, what the apostle Paul wrote in his *Letter to the Romans*, Chapter 8: «creation awaits with eager expectation the revelation of the children of God; (...) in hope that creation itself would be set free from slavery to corruption and share in the glorious freedom of the children of God» (19–21). Freedom—in the reciprocal gift of self disclosed by grace—is the epiphany of glory, that radiant light in which everything exists ontologically and will appear for what it is in truth.

That theology which is more sensitive and attentive to the issues raised about the universe by the new scientific perspectives (theory of relativity, uncertainty principle, quantum theory), in the last century, has opened up a dialogue which has tended to overcome those ideological barriers on both sides that had previously caused centuries of indifference, if not outright hostility. The rediscovery of what has been defined as "specifically" Christian has been decisive in this respect (see H.U. von Balthasar, J. Ratzinger, K. Hemmerle, W. Kasper); that is, the revelation in Christ of the Being of God as Trinitarian love. Teilhard de Chardin was a pioneer in his attempt to show the integrability, or rather, the convergence, between the evolutionary vision of the universe proposed by modern science and the Christocentric and Christo-finalistic interpretation proposed by New Testament faith. Following his endeavors, theology from all the Christian traditions came up with important studies in the effort to reinterpret the principle of creation in Trinitarian terms, in meditative comparison with the most widely accredited scientific results. Representatives are found in J. Moltmann and W. Pannenberg, from the theology of the reform; J. Polkinghorne, from the Anglican world; A. Ganoczy, J.-M. Maldamé, and A. Gesché, from Catholic theology; D. Staniloae and, ahead of his time, S. Bulgakov from Orthodox theology. An interesting reading is offered in cosmological terms by what has been called *process theology*, which refers to the writings of A.N. Whitehead. Two topics of particular interest for the dialogue between theology and science are inspired by interpreting the theological concept of creation within the horizon of truth that articulates the sense of being according to the logic described in the connection between grace, freedom, and creation.[22]

(a) The first concerns the form of comprehension of that specific relation between God and the world which is expressed by the concept of creation. Two ideas are central to the classical vision: the fundamental concept of *ex nihilo* and the secondary accidental concept of cause and effect. Yet when the event of Jesus Christ is taken up as the key of interpretation of the relation of creation between God and the world, the resulting paradigm that expresses this relation changes. It can no longer be one of cause and effect, but rather a paradigm of the relation between grace and freedom, which is love and thus gift of self: in God Himself between the Father and the Son in the Spirit, and in Him, with creation, as what is other than Himself. In this sense, a form of relation is signified whereby its coming from and depending on God loses every deterministic connotation of cause and effect, and whereby otherness not only implies identity but also

---

[22]For example, see Ernan McMullin, *Natural Science and Belief in a Creator: Historical Notes*, in Robert John Russell—William R. Stoeger—George V. Coyne (Edd.), *Physics, Philosophy and Theology: A Common Quest for Understanding* (Vatican City: Vatican Observatory 1988, pp. 47–79; *Cosmos ad Creation. Theology and Science in Consonance,* Ed. by Ted Peters, Abingdon Press, Nashville 1989; Italian Theological Association, *Futuro del cosmo e futuro dell'uomo,* Ed. by Saturnino Muratore, Ed. Messaggero, Padova 1995; Ignazio Sanna, *Fede, scienza e fine del mondo,* Queriniana, Brescia 1996; Jacques Fantino, *La rencontre entre science et théologie,* in "Revue des sciences religieuses", 71/1 (1997), pp. 60–78; Jean-Michel Maldamé, *Science et foi, conditions nouvelles du dialogue,* in "Revue Thomiste", 97/3 (1997), pp. 525–562; Giovanni Prodi—Maurizio Malaguti (Eds.), *Memoria dell'origine,* Quaderni Sefir 2 (Rome: Pontificia Università Lateranense—Mursia, 2001).

autonomy. It will of course be necessary to distinguish—as Christian doctrine does—between the level of the Trinity in itself, where there is a co-origin of the Father and of the Son-*Lógos* (Jn 1:1) in the Spirit, and the level of creation. Using a spatial image to describe both the distinction and the relation between these two levels in a figurative manner, we can say that, through the eternal Word/Son in the Spirit, the Father places the world "outside himself". Yet, having said this, the relation between God and the world must be considered, not only as being modelled upon the relation between the Father and the Son/Word in the Spirit, but also as being pre-formed by looking at the latter relation, since it is called upon to extend such a relation to the creatural level. According to Christian revelation, once the Son/Word has become man, it lives as a creature the same relationship it has always had with the Father in God. Thus it both reveals and realizes the intrinsic sense, dynamic, and finality of created being: to become—as Christian tradition says—son in the Son. Certainly, this has first of all an anthropological significance: it expresses the identity/vocation of the human being. But, through mankind, it expresses the identity/vocation of all creation (as we have seen was affirmed by Paul in his *Letter to the Romans*).

The Trinitarian principle of a relation between God and the world in terms of gratuitousness and freedom offers a harmonious paradigm for understanding the meaning of the transcendence and/or immanence of God with respect to the world. Though apparently plausible, the two major models that have in fact prevailed and still prevail in defining this relation are really not fully satisfactory for an unbiased comprehension of the Universe, nor as epistemological criteria for the interpretation of reality constituted by the principle of creation. The first model is that of the transcendence of God with respect to the world, without any immanence of Him in the world. Following a scheme of exteriority and even separation, this model cannot avoid generating problems for theological, metaphysical, and even cosmological interpretation. The second model is instead that of the immanence of God within the world, which often ends up negating the real otherness of God as it identifies Him with the world, in different forms. This model raises many difficulties, some of them even in opposition with each other, on the theological, metaphysical, and cosmological levels. The paradigm suggested by the Trinitarian perspective rethinks the abstract and basically dualistic (and therefore, ultimately excluding or unifying) contraposition of the transcendence and/or immanence of God with regard to the world. The paradigm thus offers an understanding of transcendence in a way that does not exclude a specific form of immanence, and of an immanence that presupposes and safeguards transcendence. It could be said that the transcendence of God is so transcendent that it expresses itself in the most perfect immanence in creation.

Using a term with a long history in philosophy, cosmology, and theology, some authors speak of the *pericoresi* (περιχώρησις) between God and the world, that is, of the reciprocal "indwelling" of one in the other, which requires and expresses their mutual otherness and distinction. So it is not by happenstance that we find the term first used with a cosmological meaning by the Greek

philosopher Anaxagoras of Clazomenae,[23] and later by
the Stoic philosophers, to express the intrinsic correlation of each reality with every other in the harmony of
the single cosmos. In theology, the same term would
be used by John Damascene (at the end of the Patristic
period) to express the mutual interiority in the distinction between the divine nature and the human nature in
Jesus Christ "without confusion and without separation",
as proclaimed by the profession of faith at the Council of
Chalcedon in 451. Theology then used the term to desig-
nate the relation of mutual indwelling between the Father,
the Son/Word, and the Holy Spirit in God the Trinity.
Such language has become of particular relevance in our
times to express in a satisfactory way the relation between
God and the world in light of a Trinitarian interpretation
of the christological event.

(b) Within this perspective, we find a second topic whose
formulation attempts to respond to a second question. Though typically theological, this question is
not without significance for cosmology and even scientific investigation. It involves the theological truth
according to which God/Father creates through the
Son/Word and the Holy Spirit. Irenaeus of Lyon provides a suggestive image here, namely that the Son/
Word and the Holy Spirit are like "the two hands" of
the Father who gives form and life to creation. This
theological truth, so clearly affirmed by Scripture,
was expressed by Scholastic theology which used

---

[23]*Fragment* 12; Hermann Diels—Walther Kranz, *Die Fragmente der
Vorsokratiker*, vol. II, Weidmannsche Verlagsbuchhandlung, Zürich/Berlin
1964, vol. 2, p. 38.

Aristotelian language to say that the incarnate Son/ Word is both the exemplar cause and final cause of creation. Likewise, the Holy Spirit is, in some way, its quasi-formal cause (K. Rahner). Such a definition emphasizes that, just as in God other is the subsisting of the Son/Word and other is the subsisting of the Holy Spirit, so it is in creation, due to the presence and work of both. In fact, in classical theology, because of the rather static and predefined conception of cosmology, almost the whole discourse ended by illustrating the role of the Word. Thus, in conformity with the prevailing Christological perspective in Western Christianity at that time, very little or no space was dedicated to the Holy Spirit, that is, the principle of life, of dynamism, of relation, of newness. Along these lines, the evolutionary and relational vision of the universe accredited to the contemporary sciences led to a renewed cosmological theory in general, and this also stimulated a rediscovery of the pneumatological dimension of the principle of creation. At the same time, a decisive renewal has been underway in pneumatology during the last few decades, in all areas of theological reflection about the Holy Spirit. The proposal offered by W. Pannenberg goes in this direction. According to him, the God/ world relation that occurs through the action *ab extra* of the Spirit of God, as the relation between the Father and the Son that gives movement, energy, and life to all creatures, can find a model of interpretation in the "magnetic field" (developed in the theories of physics, starting with M. Faraday).[24]

---

[24]Wolfhart Pannenberg, *Systematic Theology*, Trans. by Geoffrey W. Bromiley (Grand Rapids [MI]: Eerdmans, 1991).

The grammar of the Trinitarian relation—as grace and freedom—between the Father, the Son, and the Holy Spirit, can therefore also offer insights that may help us to understand the genesis and dynamic structure of created reality. Indeed, the latter testifies first of all that there is an origin/beginning from which everything springs (in both a metaphysical and a temporal sense) and that, as such, it is certainly inaccessible to any scientific method. Therefore, it is impossible for any naïve and dangerous form of concordism to occur between scientific results and what is revealed, since they remain on two different levels. Created reality is structured according to a plan and a dynamic that manifest their intelligibility, taking on a distinct form from one time to the next. This form, on the other hand, is neither isolated nor static, because it is the result of multiple relationality (at various levels, e.g., subatomic, atomic, chemical, biological, psychological) both within itself and in the broader context of evolution and expansion in which it is located. In this way a dynamic self-transcendence is realized, which at each moment of time involves abandoning the previous form and equilibrium in order to access new and ever more complex forms and equilibria. The greater stability of the latter does not contradict, but rather precedes and in turn renders possible the passage to other figures and higher levels. This opens the way for numerous avenues of research, many of which have not yet been explored.

What is essential, in my opinion, is not to render absolute any of the points of view involved, whether they be theological or philosophical or scientific. Rather it is important to allow them to interact, fully respecting the specific formality of each and their particular level of operation. Such interaction is based on attentive and unbiased listening to the respective reasons of each. Even this relation, in the end, is a question of grace and freedom.

Bibliography

H. U. von Balthasar, *Theo-drama: Theological Dramatic Theory*, 5 vols. (Ignatius Press, San Francisco, 1988–1998)

S. Bulgakov, *The Bride of the Lamb*, Trans. by B. Jakim (W.B. Eerdmans Publishing Company—T&T Clark, Grand Rapids (MI)—Edinburgh, 2002)

A. Ganoczy, *Der dreieinige Schöpfer: Trinitätstheologie und Synergie* (Darmstadt, Wissenschaftliche Buchgesellschaft, 2001)

A. Gesché, *Dieu pour penser*, 7 vols. (Editions du Cerf, Paris, 1993–2003)

K. Hemmerle, *Thesen zu einer trinitarischen Ontologie (1976)* (Johannes Verlag Einsiedeln, Freiburg, 1992)

W. Kasper, *An Introduction to Christian Faith*, Trans. by V. Green (Paulist Press, New York, NY, 1980)

B. Lonergan, *Method in Theology* (Lonergan Research Institute–University of Toronto Press, Toronto, 1990)

J. Maritain, *Distinguish to Unite, or, The Degrees of Knowledge*. Trans. by G.B. Phelan (Notre Dame (IN), University of Notre Dame Press, 1995)

J. Moltmann, *The Trinity and the Kingdom: The Doctrine of God*, Trans. by M. Kohl (Harper & Row, San Francisco, CA, 1981)

J.J. O'Donnell, *The Mystery of the Triune God* (Paulist Press, New York, 1989)

W. Pannenberg, *Systematic Theology*, Trans. by G.W. Bromiley (Grand Rapids (MI), Eerdmans, 1991)

L. Pareyson, *Ontologia della libertà. Il male e la sofferenza* (Einaudi, Torino, 1995)

J. Polkinghorne, *Science and the Trinity: The Christian Encounter with Reality* (Yale University Press, New Haven, CT, 2004)

J. Ratzinger, *Introduction to Christianity*, Trans. by J.R. Foster (Ignatius Press, San Francisco, CA, 2000)

P. Ricoeur, *The Course of Recognition*, Trans. by D. Pellauer (Harvard University Press, Cambridge (MA), 2007)

A. Rosmini, *Theosophy*, 3 vols., Trans. by D. Cleary, T. Watson (Rosmini House, Durham, 2007–2011)

D. Staniloae, *The Holy Trinity: In the Beginning There Was Love*, Trans. by R. Clark (Holy Cross Orthodox Press, Brookline, MA, 2012)

Printed in the United States
By Bookmasters